THE ART OF ENCOURAGING INVENTION:

A New Approach to Government Innovation Policy

The Industrial Institute for Economic and Social Research

The Art of Encouraging Invention:

A New Approach to Government Innovation Policy

by
Stefan Fölster

"Results? Why, man, I have gotten a lot of results. I know several thousand things that won't work."

Thomas Alva Edison

Distribution: Almqvist & Wiksell International, Stockholm

© 1991 Stefan Fölster and The Industrial Institute for Economic and Social Research
ISBN 91-7204-365-2
Gotab 93253, Stockholm 1991

Preface

It is now widely recognized that a country's ability to adopt and develop new technology is vital to its long-term economic growth. Public policy can stimulate or stifle technological capability in many ways. The educational system, the tax system, and even macroeconomic policies can be decisive in creating an environment favorable to innovation. The Industrial Institute for Economic and Social Research has a long tradition of supporting economic research in these areas.

This book fills a gap in which economic analysis has been relatively sparse. It examines the public subsidies commonly used in many countries to stimulate innovation in private firms.

The empirical and theoretical research presented in this book indicates that the subsidy systems that are most common today are not the most effective ones. In many cases it is questionable whether these programs are worthwhile from a social point of view. This is exacerbated by the fact that subsidizing agencies usually do not measure the effectiveness of their programs.

Subsidy programs could often be much more effective if better subsidy forms and evaluation methods were chosen. I am confident that the research in this book will contribute to the design of policies that more vigorously stimulate technological progress.

Stockholm in April 1991

Gunnar Eliasson

Table of contents

Part one: Introduction

1	**What is wrong with innovation policy?**	9
	1.1 Introduction	9
	1.2 The state of uncertainty	10
	1.3 Problems with current innovation policies	11
	1.4 Policy recommendations	13
	1.5 The organization of this book	14

Part two: An assessment of current policies   17

2	**The pattern of current policy**	17
	2.1 Introduction	17
	2.2 The cost of innovation policy	17
	2.3 Innovation policy tools	22
	2.4 Subsidy instruments	24
3	**Why public innovation policies?**	26
4	**Empirical evaluation of the effectiveness of current policies**	31
	4.1 Introduction	31
	4.2 Previous empirical studies	31
	4.2.1 Introduction	31
	4.2.2 General subsidies	33
	4.2.3 Selective subsidies	35
	4.3 The Swedish survey study	38
	4.4 An experiment on project selection	40
	4.5 Other innovation policy issues	42

Part three: Reforming innovation policy   47

5	**The politics of innovation policy**	48
	5.1 Introduction	48
	5.2 The formation of innovation policy	48
	5.3 An independent policy evaluation	52
6	**Performance measurement**	55
	6.1 Introduction	55
	6.2 Experimental policy evaluation	55
	6.3 Project evaluation	59
7	**The custom design of subsidy systems**	62
	7.1 Introduction	62

	7.2	Theoretical principles	62
	7.3	Practical application of optimal subsidies	65
	7.4	Subsidy targets and rules	67
8	**Summary of policy conclusions**		70

Part four: Selected studies — 75

9	**The efficiency of innovation subsidies**		75
	9.1	Introduction	75
	9.2	The survey design	77
	9.3	General judgements of policy effectiveness	79
	9.4	Judgements of policy effects on specific projects	80
Appendix			86

10	**Performance measurement**		87
	10.1	Introduction	87
	10.2	A simple scheme for estimating social value	90
	10.3	The computerized cost-benefit program EPRO	98

11	**An experimental test of cost-benefit methods**		105
	11.1	Introduction	105
	11.2	A laboratory experiment	107
	11.3	A real life experiment	109

12	**The incentive subsidy**		116
	12.1	The impossibility of incentive compatibility	116
	12.2	The incentive subsidy	119
Appendix			127

Bibliography — 129

Part one: Introduction

1 What is wrong with innovation policy?

1.1 Introduction

A designer of public innovation policy tends to find himself in the same position as a blindfolded brainsurgeon. After an operation he can rarely surmise what incision, if any, caused the patient's recovery – or demise. The Gordian knot of innovation policies is to know what innovations would have occurred even without a policy. The brainsurgeon, even though unable to observe the direct consequences of his actions, has the advantage of being able to draw upon known facts and reliable medical research. A designer of innovation policy is less fortunate in this respect and operates in some respects at the same level as a seventeenth century quack.

The uncertainty about the consequences of innovation policies is widely admitted by experts, although by some only in private. In public, innovation policies are frequently hailed as vital for economic competitiveness. Hand in hand with disillusionement over demand policies as a remedy of slow growth governments all over the world have intensified efforts to kick-start ailing industries with aid to R & D and diffusion of technology.

How successful the proponents of innovation policies have been in presenting their case is evident in the large and growing sums spent on them. In the OECD countries as a whole total public R & D expenditure comprises over 1% of GNP. It has been rising at a real rate of over 3% since the mid-seventies (OECD, 1986) – a period where governments have otherwise been more disposed towards slashing budgets. This public largesse has benefited most types of innovation policy, everything from direct subsidization of firms' R & D and R & D tax relief to government research institutes and, to varying degrees, even university research.

The purpose of this book is to collate the empirical and theoretical evidence concerning the effectiveness of innovation policies that involve a public subsidy to private firms or innovators. This leaves out a series of wider innovation policy questions such as regulatory questions, competition and trade questions, as well as procurement policy.

A number of new empirical studies are presented here. They suggest that the most common innovation policies are not the most effective ones.

1.2 The state of uncertainty

Anyone interested in investigating what is known about the effectiveness of public R & D subsidies may reasonably begin his quest by asking how large the total sum of R & D subsidies is in any given country. He will then make the sobering discovery that in most countries even this simple item of information cannot be attained. The OECD, having recently initiated a programme to map public subsidies to industry, is complaining bitterly that governments all too often hide such information and that there are many arrangements, such as loan guarantees, for which it is a challenge to deduce what the subsidy component is.

That governments conceal subsidies they give is perhaps understandable. Subsidies can be interpreted as trade barriers and invoke reprisals by other countries.[1] What is more surprising is that the surging government funding of R & D has sparked relatively few thorough evaluations of the innovation policy instruments at hand. What is commonly sold as "policy evaluation" is too often nothing more than a survey of the informed opinions of representatives for firms, public agencies, or consultants.

One obvious reason for the scarcity of evaluations that pass even a weak test of scientific objectivity is that such investigations face cumbersome methodological hurdles. For example, extrapolating what a firm might have done, had it not received a subsidy, is a task strewn with opportunities for generating biased conclusions.

A second reason is that economists have traditionally eshewed many areas concerning efficiency in the public sector. When such issues have been raised they have usually been treated in an abstract way with little attention paid to the practical constraints and administrative issues that a public official needs to ponder before instituting a policy.

A third reason for the absence of thorough evaluations is that the public agencies doling out funds often do not receive directives or incentives for investigating the efficiency of the tools they handle. At times they are forced merely to carry out what governments have, out of political expediency, already decided upon. Other times, they protect their own territory by resisting change as well as inspection by outsiders.

The lack of attention to the effects of innovation policy is serious because it can gravely harm the competitiveness of an economy. The

---

[1] In the European community for example subsidies to industry are generally forbidden under articles 92–94 of the Rome treaty. Exceptions are made however for certain types of R & D subsidies, subsidies to less developed regions, and a few other types of subsidies.

wrong kind of support for R & D is not just a way of harmlessly throwing away tax payer's money, but it may actually stifle innovation. The danger is perhaps greater than that of waste in many other areas because the conditions that are conducive to creativity and innovation are extremely fragile. For example, subsidies directed at one type of research may primarily have the effect of breaking up productive research teams elsewhere that may never again produce results of the same quality. Equally, the talents of a generation's brightest students can be lost permanently if they are guided into the wrong fields.

A misguided policy may easily draw scarce talent into unproductive fields at the price of neglecting promising opportunities. Anyone doubting policy maker's capacity for error should be reminded of Lord Kelvin's proclamation, while serving as president of the Royal society, that radio had no future and that X-rays were a hoax. Equally embarassing with the benefit of hindsight, the French marshal Ferdinand Foch dismissed airplanes as interesting toys of no military value as late as 1911. On a more recent note, the Swedish minister of trade, Gunnar Lange, told Volvo 30 years ago that trying to sell cars to the U.S. was as stupid as trying to sell refrigerators to the eskimos. Exports to the U.S. later became Volvo's most profitable line of business.

While R & D subsidies can turn out to be expensive mistakes it is also true that a missed opportunity may leave a country at a strategic disadvantage from which it may never recover. One such example is the development of the transistor after its invention in 1948. At the time European firms were the equal of American ones in the development of electronic components. Yet few firms anywhere were willing to invest in transistors without some kind of government support. In the U.S. firms received this support, achieved a headstart, and have maintained it ever since.[2]

1.3 Problems with current innovation policies

The problem of designing innovation policy is captured by three questions:

1. Who should one apply innovation policy to?
2. What instruments should one use?
3. How should one evaluate the success of innovation policies?

A successful solution to the third question would presumably provide the information required to answer the first and second question. The third question also raises the issue of whether an innovation policy is

---

[2] See Schnee (1978) for a fine account of this development.

worthwhile at all from a social viewpoint. In chapter 3 we discuss the theoretical arguments for and against innovation policies. In chapter 4 the empirical literature on the success of innovation policies is discussed. Since the issue of whether R & D subsidies are worthwhile is so central to this book, however, we will begin immediately with an illustration of how to tackle this question.

We have already lamented the fact that policy evaluations are rare today. The scant body of extant empirical tests, however, is not kind to many commonly used policies. Innovation subsidies are frequently found to have but a small effect on firms' R & D programmes. For example, some rather typical studies conclude that grants to private R & D projects increase the firms' R & D spending by roughly 40% of the value of the subsidy. This means that firms pocket 60% of the subsidy which they receive for projects that they would have conducted anyhow. Sometimes such a result is interpreted to imply that the additional R & D generated must have a social return of at least 250% in order to recuperate the cost of the subsidy to society. This interpretation is, however, misleading since the subsidy is not in its entirety a cost to society; it is only a transfer from tax payers to firms.

A more correct interpretation of such an empirical result is illustrated in the insert below. The insert contains a hypothetical cost-benefit calculation for an innovation subsidy. The actual numbers used are intended only as an illustration but they reflect results of empirical studies that are discussed in detail in chapter 4.

**The Social Value of an Innovation subsidy:
An Illustration based on Empirical Results**

Assume that a subsidy of $ 100.- is given to a firm. Suppose further that 40% of that is used for new R & D that the firm would not have conducted without the subsidy.

The social cost of the subsidy consists of subsidy administration costs and the efficiency loss that occurs when public funds must be raised via taxes. For the sake of illustration, suppose that the social cost of granting the subsidy is 30% of the value of the subsidy. This figure is quite in line with the findings of empirical investigations of the efficiency loss due to taxation (e.g. Hansson, 1984).

This implies that inducing new R & D worth 40% of the value of the subsidy involves a social loss corresponding to 30% of the value of the subsidy. This can be expressed in the following table:

New R & D generated by subsidy	40
Efficiency loss of taxation	30
Social cost	70
Social value required for break-even	70
Required rate of social return on new R & D	75%

The required rate of social return on new R & D can be compared with the actual return to industrial R & D. Empirical studies of the social return of industrial R & D do find rather high returns, sometimes on the order of 100% (e.g. Bernstein, 1989). These studies refer, however, to average R & D projects. In contrast, the new R & D generated by subsidies is marginal. Firms would not have conducted it without the subsidy. Therefore it must be assumed to yield considerably lower private profits than the average R & D project. To the extent that private profits and social values are correlated it must therefore be assumed that even the social return is considerably lower for marginal projects.

That many innovation policies do not perform better in empirical tests is perhaps more understandable when one considers how decisions are commonly made today about who to apply innovation policy to and what instruments to use.

Case studies indicate that innovation policies often are channelled in the wrong direction to satisfy political interests. In fact, a common complaint of public initiatives is that they are frequently misdirected or support technologies too late when the exciting inventions in a field already have been made.

Even if one were to disregard the empirical tests of the efficiency of subsidy instruments we argue in this book that commonly used subsidy instruments are not well conceived from a theoretical standpoint. A closer inspection of current subsidy systems reveals that many are implemented in a way that virtually guarantees inefficiency. Agencies distributing subsidies frequently are not in a position to judge whether the project that a firm seeks subsidies for would have been conducted anyhow, or even whether it is socially valuable. Even when these agencies are well endowed with technical know-how they can rarely match the firm's inside information on market potentials. The firms, in turn, have incentives to apply primarily with projects they would have conducted anyway, pocketing the subsidy as a pure gift. In doing so firms are often forced to exaggerate a project's social value and to present a project as though it would not be conducted without the subsidy.

Further, subsidizing agencies often have a poor understanding of the circumstances under which a subsidy creates social value. We present an experiment among subsidy administrators that supports this claim.

1.4 Policy recommendations

Better policies are based on better methods for selecting projects that are worth subsidizing. It is shown in this book that some types of R & D subsidy forms are superior to others in the sense that they make it in the firms' best interests to accept government support for the best research opportunities that they would not have exploited of own accord.

A more detailed summary of policy conclusions is provided in chapter 8. Briefly, the policy conclusions fall into three categories:

1. First, they concern the politics of innovation policy. We find neither government departments nor the subsidy-granting agencies are well suited to evaluate the efficiency of innovation policies. Independent academics on the other hand have usually not been given the support and access to data needed to conduct detailed studies. We suggest that evaluation of innovation policies should be entrusted to a fairly independent agency with enough clout to ensure cooperation of the subsidy-dispersing agencies.

2. Second, a more professional approach to evaluating policy success and selecting criteria for project selection is needed. Policies should more often be designed as experiments that yield information about their efficiency. We discuss the methodological issues of policy evaluation at some length in this book.

3. Third, subsidies must be given in a way that provides correct incentives. To some extent incentives can be designed in a way that considerably reduces the government expense of achieving a certain policy objective. In many situations, however, it is found that current subsidies can be replaced by equity capital investments. Empirical results indicate that such investments can stimulate more new R & D than corresponding subsidies. These investments can even be funnelled through private investment or venture capital firms.

1.5 The organization of this book

The primary focus of this book is on subsidy systems that affect industrial firms, even though other issues such as patents, government research institutes, and direct government R & D coordination are touched upon. More indirect measures, such as support for education and university research, are not treated here. This selection reflects the fact that much attention has been lavished on the analysis of education systems, while the literature on industrial innovation policies so far has been sparse and mainly descriptive.

Parts one to three consist of an easily accessible overview of current policies and suggestions for policy review. Part four consists of detailed reports of the empirical and theoretical work that many of the conclusions are based upon.

Part two, comprising chapters 2 to 4, is concerned with an assessment of current policies. Chapter 2 outlines the composition of current R & D policies in OECD countries. Chapter 3 summarizes the economic arguments for innovation policies. This defines the aims of policy instruments. Chapter 4 assesses the effectiveness of current policies based on previous empirical work and our own empirical comparisons of subsidy instruments.

Part three, comprising chapters 5 to 8, is concerned with policy conclusions. The innovation policy apparatus is analysed here and suggestions for reforms are given. Chapter 5 points to some problems with the political process that steers innovation policy. Chapter 6 discusses ways of measuring policy performance. Chapter 7 presents the principles that make for efficient subsidy instruments. Finally, chapter 8 summarizes the policy conclusions.

Part four consists of a collection of papers that describe in greater detail the empirical and theoretical work on which this book is based.

Part two: An assessment of current policies

2 The pattern of current policy

2.1 Introduction

This chapter provides a brief overview over the extent and character of innovation policy in selected countries. We begin with a comparison of the total costs of innovation policies based on the rudimentary statistics that are available. Then we review some surveys on the choice of innovation policy tools. Finally, we focus on the specific subsidy instruments that can be used.

2.2 The cost of innovation policy

The total level of public funding for R & D can be seen in Table 2.1. Interestingly a number of the countries that are among the highest in terms of total R & D as a percentage of GNP, such as Sweden, Germany, and Japan, do not have particularly large public outlays for R & D. The U.S. in contrast has a high level of total R & D, but much of that consists of public R & D expenditure, often toward military ends.

Figures 2.1 and 2.2 shows that although governments have increased their R & D efforts, the business sector boasts even larger

Table 2.1 Total R & D and total public expenditure on R & D as percent of GNP

	Total R&D 1987	Public Exp 1987	Public Exp 1979
Belgium	1.5	0.70	0.59
Denmark	1.3	0.60	0.48
FR Germany	2.7	0.97	1.13
Greece	0.3	0.22	0.19
France	2.3	1.19	1.09
Ireland	0.8	0.39	0.53
Italy	1.1	0.55	0.39
Netherlands	2.1	0.98	0.96
U.K.	2.3	0.98	1.07
Sweden	2.8	0.95	1.00
U.S.	2.8	1.40	1.32
Japan	2.6	0.50	0.65

Source: OECD, UNESCO

Figure 2.1 Growth of Public R & D Funding

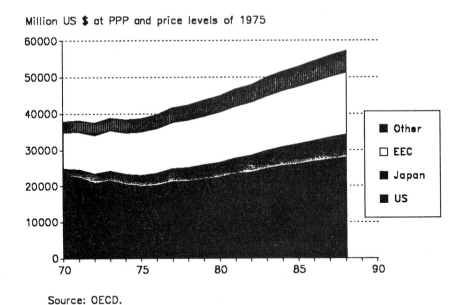

Source: OECD.

Figure 2.2 Growth of Private R & D Funding

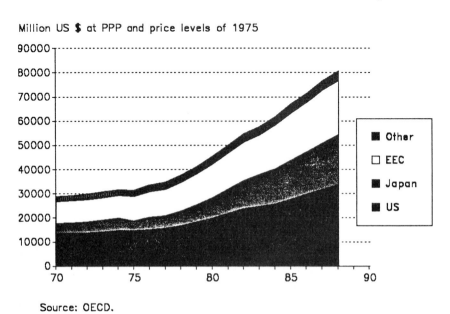

Source: OECD.

increases. Thus the enterprise sector stands for a larger segment of the total R & D conducted now than in the seventies.

Total public R & D expenditure includes research performed at universities and within the public sector. Unfortunately, precise statistics on government subsidies to private innovation are not available. The OECD has begun to collect such statistics. The series is not complete, however, and there remain a host of definitional problems to be solved. For example, it is unclear what the subsidy component of loan guarantees and public procurement contracts is. As another example it hard to define whether tax deductions for R & D expenses should be considered to be subsidies or just a regular part of the tax system.

For the time being, the best one can do is to approximate the desired measure. One available measure is the fraction of public R & D spending which is directed toward industrial development, agriculture, energy, infrastructure and civil space. The resulting measure, shown in Table 2.2, still contains a number of items that one would not necessarily consider to be innovation policy. On the other hand, some items that probably should be considered to be innovation policy are not included, such as tax relief for R & D and the advantage firms receive from defense research.

Table 2.2 Public R & D spending on industrial development, agriculture, energy, infrastructure, and civil space, in percent of GNP

	1987
Belgium	0.28
Denmark	0.20
FR Germany	0.33
Greece	0.09
France	0.36
Ireland	0.20
Italy	0.25
Netherlands	0.27
U.K.	0.19
Sweden	0.18
U.S.	0.20
Japan	–

Source: OECD

Another way of approximating the level of public subsidies to private innovation efforts is to look at where the publicly funded R & D is performed. This is shown in Figure 2.3. Alas defense R & D expenditure is included in this figure, again providing a somewhat distorted picture of the extent of innovation policy expenditures. Also, some

Figure 2.3 Recipients of public R & D funds, 1987.

Source: OECD.

subsidy instruments such as tax credits are not included. A clear pattern is that the countries that perform much defense R & D also divert much of public R & D expenditure to the private sector, where defense R & D usually is performed.

A first, heuristic, test of innovation policies might be to look for correlations between measures of R & D and economic performance on a macroeconomic scale. If one found no correlation between R & D performed and economic growth then one might suspect that innovation policies designed to raise R & D are not worthwhile.[3]

The amount of total spending on R & D is not obviously correlated with the growth of GNP or of industrial production or the amount of private expenditure on R & D, or even with increases in productivity. In Figure 2.4a we show a narrower measure of R & D and its relation to the average annual GNP growth between 1979 and 1984. All OECD countries are plotted in this diagram. Clearly there is no particularly convincing relationship.

---

[3] Although an alternative interpretation might be that those countries that fall behind in economic growth invest more in R & D in order to "catch up".

If one chooses an even narrower measure however the fit is better. Figure 2.4b shows the level of industrial R & D plotted against average annual GNP growth. In this diagram the fit seems much better. The main outlier with a high level of R & D but low growth is Sweden. A possible explanation is that Sweden is the home base for an exceptionally large number of multinational companies in relation to its size. These companies have often concentrated their R & D activities to Sweden but produce elsewhere.

Figure 2.4a. Civilian Research and Development and GNP growth 1979–1984.

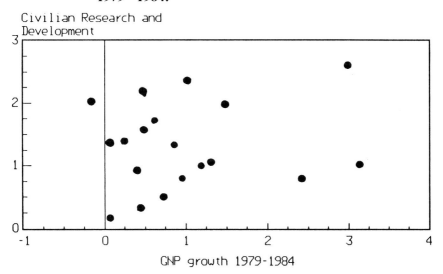

Figure 2.4b. Industrial Research and Development and GNP growth 1979–1984.

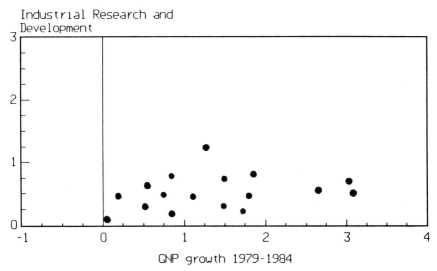

Figure 2.4c. Public expenditure for industrial R & D and GNP growth 1979-84.

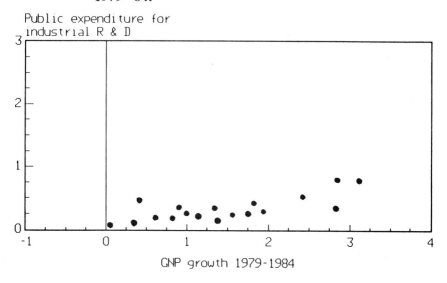

Finally, Figure 2.4c shows the relationship between public expenditure on industrial R & D and GNP growth. Again there is no discernable relationship. This is not surprising since there are several mechanisms at work simultaneously, apart from the presumed increase in growth as a consequence of raised public investment. For one, as we argue in this book, some types of public R & D may not be very effective. Further, countries' willingness to support private R & D may rise with deteriorating economic performance. This would indicate a negative correlation between growth and public investment. Also, the industrial structure differs widely between countries. For example, in Japan firms seem to invest large amounts in R & D out of their own initiative. Thus the government may feel less compelled to invest itself.

2.3 Innovation policy tools

Adopting a broad definition of innovation policy the policy types can be classified as in Table 2.3.

Table 2.3 Components of innovation policy

1. Subsidies including Tax Credits
2. Education
3. University and Government R & D
4. Management of the Public Sector, including Government Procurement and Public Services
5. Laws and Regulations including Patent Law
6. Foreign Trade Policy

Table 2.4 Analysis of policy recommendations by type of tool, in percent of country total

	Japan	Sweden	U.K.	U.S.	Netherlands
Rothwell & Zegveld, 1981:					
1. Subsidies and tax credits to firms	11	20	33	18	15
2. Education	5	37	11	3	12
3. University and government R & D	41	13	14	4	22
4. Management of the public sector	35	23	36	22	32
5. Laws and regulations	0	3	0	49	15
6. Foreign trade policy	6	3	6	3	3
Fölster, 1988:					
1. Subsidies and tax credits to firms	7	33	25	23	18
2. Education	7	21	19	15	11
3. University and government R & D	38	16	12	18	25
4. Management of the public sector	33	24	31	19	29
5. Laws and regulations	4	2	3	16	14
6. Foreign trade policy	9	5	10	9	3

This book is concerned mostly with the first category. Before focusing on subsidies, however, it is interesting to see how different countries have quite different views of how innovation policy should be conducted. To study this we have counted the number of policy recommendations in a number of public innovation policy statements from 1987.[4] In doing so we have repeated an earlier similar analysis by Rothwell and Zegveld (1981). They use a different classification of policy types. Reinterpreting their classification in terms of ours yields the pattern shown in Table 2.4.

This comparison shows a clear difference between the U.S. and other countries. If one can interpret the difference between the studies as changes beween the years 1981 and 1988 then it appears that planned innovation policy in the U.S. has become more similar to that in other countries. Primarily the preoccupation with laws and regulations as innovation policy tools seems to be diminishing in favor of subsidies and trade policy. This seems to fit the policy changes that

---

[4] The exact list of policy statements is given in Fölster (1988c).

have actually occurred. The U.S. has instituted a number of subsidy programs during the 1980s.[5]

Table 2.4 says little about what innovation policies are directed at. An interesting characterization in this respect is provided by Ergas (1987). He points out that some countries frequently subsidize large, high-tech, projects. These countries he calls the "mission-oriented" countries. Among the mission-oriented countries are France, the U.S., and Britain. Other countries, the "diffusion-oriented" countries, emphasize support to small firms and to diffusion of extant technology. The diffusion-oriented countries are Sweden, West-Germany, and the Netherlands.

2.4 Subsidy instruments

Subsidy instruments can be divided into two broad categories: General subsidies and selective subsidies. The general subsidies are

Table 2.5 A list of subsidy instruments

General subsidies
1. Tax deduction for R & D expenses
2. Tax deduction for a rise in R & D expenses
3. Personnel grant toward costs of R & D personnel

Selective non-self-financing subsidies
4. Project grants
5. Project loans at subsidized interest rates
6. Conditional loans that are repaid only if R & D is succesful
7. Loan guarantees
8. Prizes

Selective self-financing subsidies
9. Fee-based loan guarantees
10. Royalty grants. Royalty to the state is based on sales of the invention toward which the grant was applied.
11. Stock option grants. In return for an R & D grant the state receives a stock option that can be exercised if the stock value rises significantly. For large firms the stock option refers to separate venture companies set up around the respective R & D project.
12. Convertible loans. The state gives a loan that can be converted into stock if the project turns out to be a commercial success.
13. Equity investments. The state invests in venture firms either directly or indirectly via private investment companies.

---

[5] One example is the Small Business Innovation Act that was passed in 1982. This requires that every Federal Department or Agency with an extramural R & D budget of more that $100 million annually must establish a Small Business Innovation Research programme. Such a programme diverts a target proportion of 1,25% of each budget to the financing of R & D projects in firms with no more than 500 employees.

given to all firms according to certain well-defined criteria. There is no case by case selection of projects that should be subsidized. In contrast, the selective subsidies are given to projects that a subsidizing agency chooses from a list of applicants. The selective subsidies can be divided further into self-financing and non-self-financing subsidies. The self-financing subsidies contain provisions that, in principle, make it possible for the subsidy to be costless to the public purse. For example, a fee-based loan guarantee may break even, if the fees are set sufficiently high. In contrast, a pure grant system does not even have a theoretical chance of breaking even.

Table 2.5 shows a list of subsidy instruments. This list is not complete but most of the omissions are variations of instruments on the list.

All of the subsidy instruments in Table 2.5 are in use somewhere. In most countries grants and project- or conditional loans are most common.

3 Why public innovation policies?

This chapter briefly summarizes the principal theoretical arguments for and against innovation policies.

Firms spend money innovating in expectation of resulting profits. The profits are large if few other firms innovate simultaneously, allowing the firm to exploit a cost or quality advantage. Firms lose if rivals can imitate the invention and produce cheaply without having incurred research costs. Witness the microcomputer market. The first machines made millionaires because only few had the know-how to build them. Now, with the technology in everyone's hands, many firms lose on their microcomputer production.

If the inventing firm does not capture all gain to an invention then it does not have sufficient incentive to research from a social point of view (see e.g. Arrow, 1962). There will be some projects it discards even though they profit society. Imitation is not the only way a firm can be deprived of some of the social benefits of innovation. It may not be able to capture the entire consumer surplus (i.e. the users' benefits from the invention), and some of the costs of educating researchers and engineers may be lost when these succumb to the lure of rivals or of independent ventures.

To ameliorate the problems with imitation the patent office was instituted. With an invention patented a firm stands a chance of being the sole exploiter, at least during the patent's life. This raises the return to the inventing firm and its incentive to research.

Unfortunately, there are three problems with patents. First, too often they do not work.[6] Imitation, or inventing around the patent is too easy and may even be facilitated by the required disclosure of technical information during the patenting procedure. Second, when patents work the inventing firm becomes a monopolist for the invention. Monopolies are known to engender social welfare loss. Third, patents are hard to adjust to individual circumstances. If imitation is easy the length of patent life may be irrelevant. If imitation is difficult patent life can be too long inducing too many firms to research toward the same invention too fast, each trying to win the patent race.

If a patent fails to defy imitators not all is lost. Imitators face two other hurdles. First, the knowledge required to produce the invention may not be easily transferable. It may consist mostly of skills that employees have acquired. Would-be imitators must then move down a learning curve before they can compete; or they can try to lure employees from the leading firm. Second, the inventing firm can try

---

[6] This is shown in empirical studies by e.g. Mansfield et al. (1981) or Levin et al. (1987).

to keep information about inventions secret. The trouble is that secrecy is expensive for the firm and adds nothing directly to social welfare.

In sum the firm is likely to research insufficiently from a social point of view in areas where inventions are difficult to keep secret either because know-how leaks easily and patents protect poorly, or where results emerge only in the long run so that people with non-transferable knowledge may quit. Empirical investigations of the returns to industrial R & D usually find that the social return is considerably larger than the private return (e.g. Bernstein, 1989).

This is the most commonly cited reason for government intervention. The other reasons are more controversial. They can be grouped into the following categories.

Risk aversion

Firms may be too risk averse to engage in projects that are large in relation to the size of their company. Since society as a whole can spread risks better, and therefore is less risk averse, the government should then assume some of the risk of research projects or subsidize some of the riskiest projects. This argument is sometimes phrased as a failure of capital markets to provide risk capital. Institutional factors, for example, seem to play an important role in determining the extent and structure of venture capital markets.

Fölster (1988b) shows in an empirical study of Swedish firms that firms with larger profits in relation to size accepted research projects with a lower expected profitability than firms with small profits.[7] This can be interpreted as supporting the thesis that a greater risk of bancruptcy raises the degree of risk aversion in selecting R & D projects.

A common criticism of the risk aversion argument is that it is easily used as an excuse for subsidizing in a situation where it actually is uncertain whether a firm would or would not invest without a subsidy. More about this problem will be said later. Note also that taken literally, the risk aversion argument implies that many kinds of private investments should be subsidized, not just R & D investments.

When banks lend money they may demand higher interest rates when the risk of default is higher. Sometimes it may by better for banks however to retain a low interest rate and ration credit (Stiglitz & Weiss, 1981). The reason for this is that banks cannot perfectly observe individual firms' default risk. If they raise interest rates they are likely to lose those customers with the lowest default risk first.

---

[7] A few firms however pursued a policy of maintaining a high profit rate by only investing in profitable projects. Thus different firms seem to pursue different strategies.

In the innovation policy debate it has often been argued that small firms do not have sufficient access to the capital in order to expand.

Slow diffusion of technology

Technology adoption often is a process of firms imitating what they see other firms do. When one firm adopts a technology it thereby provides an external benefit by showing other firms whether the technology is profitable. Therefore public measures to speed up the spread of a technology may sometimes be justified.

Two qualifications of this argument are in order. First, what looks like R & D at the firm level is often nothing but adaptation of extant technology. Therefore this argument is really quite similar to the one made above stating that firm R & D perpetrates inappropriable external effects. Second, aiding diffusion of a technology often reduces the period during which the original inventor earns profits. Thus incentives to invent may be reduced.

A thorough review of the literature on technology diffusion is found in Stoneman (1983).

Increasing returns to scale in R & D

If an industry's R & D efforts become more productive with size then there may be a case for public assistance in getting the industry up to the size where it can compete internationally. This is a version of the infant industry argument. Japan is one country that claims to have succeeded with this method. The government focuses on a new technology, forces firms to cooperate in developing it, supports them with funds, protects the home market during the early production stages, and finally pushes the newly born industry to unleash the now mature product on the world markets (see e.g. Oshima, 1984).

In the economic literature this argument is viewed with scepticism. It is argued that if there are increasing returns to scale then it should be possible for a single firm to invest on a sufficient scale if it can then operate profitably and recoup the initial investment.

For several reasons however a single firm may be prevented from investing on a large enough scale. Fölster (1988b) analyses a case where each firm's investment in R & D generates external effects. Each firm's investment opportunities are also dependent on unique knowledge or ideas. In addition there are increasing returns to scale to the combined R & D effort of all firms. Then it turns out that no firm has an incentive to invest first and, more importantly, no firm may be able to extract profits from coordinating a common R & D effort or buying out the other firms.

The common pool problem

A number of theoretical arguments have been made for the case where patents work perfectly or where, for other reasons, the winner of a race towards an invention claims the entire gain, and there is free access to the research idea. This literature has been somewhat disappointing because it turns out that very small changes in the assumptions about the game among the competitors fundamentally changes the results.

Some findings appear fairly robust however. Among these are that firms can invest excessively in R & D (e.g. Dasgupta & Stiglitz, 1980). A related point is that firms may choose research projects that are too closely correlated from a social point of view if choosing correlated projects is cheaper than choosing uncorrelated projects (Dasgupta and Maskin, 1987).

The intuition behind these results can be understood in terms of external effects. When a firm begins to research it inflicts a negative externality on other firms already researching. These firms now have a smaller chance of winning. The entering firm does not include this externality into its profit calculation. As a result the private outcomes will not in general equal the socially most efficient one.

This was a synopsis of the theoretical arguments that imply deviations of firm R & D from the socially optimal levels. The existence of inefficiency in the market for innovation does not by itself justify government intervention. In addition there must be reason to believe that a government policy can ameliorate the market failure at an acceptable cost. That places high demands on the efficiency of government policy.

Discussions of subsidies in general emphasize a political problem with subsidies. Once instituted they are difficult to abolish. For that reason it is sometimes argued that taxation may be better way of reaching a goal. For example, taxing environmental pollution may be more sensible than subsidizing cleaner technology. In this example taxation has the additional advantage of raising the cost of environmentally harmful production.

For the purpose of innovation policy the option of taxation does not always exist. Clearly one cannot tax firms in general for "non-conduct" of R & D. If one instead taxes some consequence of failure to innovate taxes may be less efficient than subsidies in effecting innovation. Consider again the example of firms failing to research toward environmentally cleaner technology. Suppose that there are two market failures. First, there is the negative externality of pollution and, second, there is a positive externality to conducting R & D. A tax on pollution then does not solve the market failure in R & D and may therefore have very different effects from a subsidy of environmental R & D. It is easy to construe situations where the subsidy is much more effective than the tax.

The costs of a subsidy consist of the administrative costs and the efficiency loss incurred when raising the subsidy funds via taxation. The benefits of a subsidy depend on how effective it is. Three problems can render R & D subsidies ineffective:

1. Projects are supported that are not socially worthwhile.
2. Projects are supported that would have been conducted anyway.
3. Projects are supported in a way that leads to less efficient conduct.

In chapter 7 we summarize the theoretical analysis of subsidy instruments designed to avoid these three problems. In chapter 12 this analysis is presented in depth.

4 Empirical evaluation of the effectiveness of current policies

4.1 Introduction

A growing number of studies have increased our understanding of the process that drives technological advance (for surveys see Kamien & Schwartz, 1982; Stoneman, 1983; Dosi, 1988). At the same time an increasing awareness of the importance of technological advance for economic growth has shifted the main focus of government policy in many countries.

Unfortunately little is known about the effects and effectiveness of government R & D policy. Most of the literature in this area concerns theoretical issues or broad strategic questions (see for example Stoneman, 1987; Dasgupta & Stoneman, 1988; Hall, 1986).

A few empirical attempts have been made at evaluating the effectiveness of innovation policy. The following chapter congregates the evidence. Section 4.2 examines previous empirical investigations concerning the effectiveness of subsidies. Although the empirical results are patchy some clear indications emerge about how not to subsidize and in what directions to search for more effective subsidy policies.

In the third section, 4.3, we summarize the results of our own investigation of the effectiveness of subsidy instruments which we somewhat unimaginatively call the Swedish study. A full account of this investigation is supplied in chapter 9. This study is, as far as we know, the only study that compares different subsidy instruments using the same sample of projects.

In the fourth section, 4.4, we summarize a classical experiment concerning the process by which subsidizing agencies select projects that receive subsidies. The experiment seems to reveal a number of cases where subsidy administrators had a poor understanding of whether a subsidy is socially worthwhile. A full account of the experiment is given in chapter 11.

In the fifth section, 4.5, we discuss the empirical evidence on a number of other issues of innovation policy, such as patents and research institutes.

4.2 Previous empirical studies

4.2.1 Introduction

In theoretical models the "efficiency" of subsidies is easily defined as the change in some social welfare function. A handful of empirical studies in fact attempt to measure the social value generated by subsidized R & D. One example is Griliches' (1957) investigation of the social value of the research toward hybrid corn.

Most empirical studies however have concentrated on the question of how much additional R & D is generated for a given cost to the public purse. This means that a clear policy conclusion can not always be drawn. Policy conclusions are undisputable only when the additional R & D generated is negligible or when the relative effect of different subsidy instruments can be compared. Measurement of additional R & D also ignores a number of efficiency aspects such as the extent to which the conduct of R & D is adversely affected by the subsidy application procedure and subsidy regulations. Nevertheless it probably captures the central element and it is empirically tractable in a wide range of circumstances. Also it is one of the major ingredients in cost-benefit studies as illustrated in chapter 1.3.

Some researchers have attempted to address the question of subsidy effectiveness by examining whether government sponsored projects led to commercialized products less often than non-subsidized projects. For example Ettlie (1982), in a study of federally sponsored innovation projects in the U.S., finds that subsidized industrial projects lead to commercialized projects more seldom than unsubsidized projects (also Allen et al., 1978).

Using an econometric approach that compares industry branshes in relation to government support, Griliches (1980), Link (1981), and Terleckyi (1980) find that the private rate of return to government-financed R & D appears far lower than that for company R & D.

It is difficult however to infer from these results that subsidies are inefficient. The aim of the subsidies is after all to support projects that firms would not otherwise conduct but that are socially valuable due to indirect influences or because they are easily imitated. Thus the projects supported by the government should probably show a lower private return.

Among the studies that attempt to measure how much additional R & D is generated by a subsidy four empirical approaches are prevalent.

The first empirical approach is the case study. Individual policies or subsidies are described based on interviews and simple descriptive data. This approach often gives a good insight into practical problems with policies and the opinions of the people involved. It is often unclear however how representative they are. A subsidy seems to have very different effects from one project to the next. How effecive a subsidy policy is depends on its average effect. It is difficult to see that case studies can say much about the average effectiveness of subsidies. In addition is is difficult for the researcher to convey how objective her conclusions are.

The second approach is an econometric analysis of industry or firm data. The basic idea is to correlate the amount of subsidy a firm receives with the R & D it conducts itself. Using various forms of regression analysis these studies often show interesting correlations. Their achilles heel is their failure to identify causal relationships. Often one suspects that the most important causal factors are not even among the measured variables. For example suppose one finds that firms that receive more subsidies also conduct more R & D. This may be because the subsidies induce R & D, or because subsidies are given to firms that conduct much R & D, or because firms that have bright ideas find it profitable to conduct a lot of R & D and also attract subsidies.

The third empirical approach consists of surveys among firm executives. For examples executives may be asked how specific decisions would have been changed in the light of certain policies; or they may be asked what consequences they believe a policy has for the industry as a whole. One criticism of surveys is that respondents may not always tell the truth. They may fear a reduction of subsidies if they respond wrongly; or they may loathe the thought of appearing dependent on government handouts? The risk for such distortions must be judged within the context of each study. In many studies respondents do not have any obvious incentive to lie. Those surveys may give a good idea of what respondents believe. Honest respondents can still be wrong. Thus a survey is more likely to be useful if one can ask questions that respondents are in a position to answer correctly. For example, it appears much more reliable to ask respondents what they would do if they received a subsidy than to ask them what effect subsidies have in their industry.

A further method of empirical investigation consists of classical experiments. Since such experiments have hardly been used we reserve a more detailed discussion of their methodological problems for chapter 6 where we make a plea for a wider use of this method as a policy evaluation tool.[8]

After this brief presentation of methodological problems we show how these techniques have been used to assess the effectiveness of general and selective subsidies.

4.2.2 General subsidies

General subsidies are those that are granted to all applicants that fulfill certain simple criteria. No case by case selection is attempted.

In many countries tax credits for R & D expenditures have been

---

[8] See for example Tassey (1985) for a description of some experiments that have been conducted.

granted usually allowing a larger deduction for an increase in R & D expenditure rather than a level of expenditure. The advantages of such a policy are that it requires little bureaucracy and that it is usually viewed favorably by private enterprise. The disadvantages are that the increase in R & D expenditure may be marginal and that much public funds go to efforts that would have been made anyhow. Bozeman and Link (1984) discuss the advantages and drawbacks of tax credits in more detail.

Empirical studies (mainly Mansfield, 1985, 1986; Mansfield & Switzer, 1985; Bernstein, 1987) in the United States, Canada, and Sweden about the effectiveness of tax credits come to surprisingly similar conclusions. Econometric results indicate that the price elasticity of R & D is around 0.3. This means that every dollar of foregone tax revenue due to the tax credit raises firms R & D expenditures by 30 cents.

Mansfield confirms this inference with a survey of firm executives. These executives were asked to estimate the effects of the tax incentives on R & D expenditures. The results were similar in all three countries with the ratio of the tax-incentive-induced increase in R & D spending to the foregone government revenue lying between 0.3 to 0.4.

Moreover there was substantial evidence that the tax incentives resulted in a considerable redefinition of preproduction activities as R & D, especially in the first few years after the introduction of the tax incentive. Such redefinition of activities is estimated to have resulted in a total increase in reported R & D expenditures of 13–14 % in one country, Sweden, over the course of a few years.[9]

Interestingly similar tax credits for capital and employment also fare poorly in empirical studies. For example Folmer and Nijkamp (1987) find that investment premiums for capital in Holland had only a slight effect.[10]

A further empirical examination of a general subsidy tests the German policy of subsidizing 25-40% of researchers' salaries in small and medium-sized firms (Meyer-Krahmer et al., 1983; Brockhoff, 1983). This programme is widely considered a success because it has been popular among firms. The authors' conclusions however paint a more ambivalent picture. In thorough interviews with firms they find that 15% of the firms used none of the grant to increase their R & D activities. Most of the remaining firms only used a fraction of the

---

[9] In Sweden there was an R & D tax allowance equaling 5% of a firm's R & D expenditure plus 30% of the increase over the previous year. This policy was terminated in 1984 due to doubts about its effectiveness.

[10] Another example is given by Bohm and Lind (1988) who perform a so called quasi-experiment on the employment effect of a general reduction of social security taxes. They find the employment effect to be minor.

grant to bolster R & D. No firm initiated a new R & D project because of the grant.

The authors also compare R & D personnel recruitment in firms that received grants with firms that did not receive grants. They find that spending on new recruitment only added up to about 15% of the total programme cost. Internal reshuffling of personnel to R & D accounted for another 30% of the programme cost, and R & D equipment for 15%. All in all 60% of the programme costs was initially estimated to have been used to strengthen R & D capacity. Later that figure was found to decline however. In this calculation a note of caution must be sounded concerning the personnel reshuffled internally. It is unclear how much of this effect involves a true switch of work performed rather than just a formal reassignment.

4.2.3 Selective subsidies

Selective subsidies are those that are given on a case by case basis depending on the judgement that the subsidizing agency makes in individual circumstances.

Case studies

In order to determine whether government subsidies are awarded too often to unworthy projects one can examine what incentives the firm and the government have in allocating subsidies. When it is in the firm's interest to receive a subsidy it will naturally try to represent the project as having a significant social value. The government bureaucrat usually has small possibilities of checking the information supplied by the firm. This problem is well documented in case studies. For example, Nelson (1982) presents case studies that reveal a pattern of relatively successful intervention in basic research, "generic" technologies, and fundamental research areas such as health and agriculture where researchers, rather than government officials, make resource allocation decisions. When governments attempted to "pick winners" and to intervene in the later stages of technological development, the results were substantially less favorable.

Further, the government bureaucrat deciding who is worthy of public funds and who does not has a self-interest that may be at variance with the common good. Two typical cases are likely to arise. One is that of a politically motivated decision. There is evidence from case studies to suggest that government support has gone to showcase projects such as pilot plants that contribute little to overall innovation but serve well as evidence of a politician's or an agency's initiative. For example Roessner (1984) shows on the basis of case studies how government R & D managers and administrators were under pressure to push technologies prematurely to commercialization status, implying highly inefficient and costly decisions. The primary source of this pressure were elected and appointed officials who sought the political rewards of short term highly visible, easily implementable programs.

The other problem with government officials' incentives is that of a government employee, responsible for distributing a certain sum of subsidy funds. The problem that arises is that his superiors have an informational disadvantage in evaluating the administrator's performance. Usually there exists no data on the expected social value of projects, or on whether the firm would have conducted this research anyway without the subsidy. The information that is most readily available is whether the supported project, after its completion, becomes a commercial success. The likelihood of a project succeeding commercially depends on two things. First, the administrator's skill in choosing winners and helping to shape the project so that it succeeds. Second the inherent riskiness of the project. The less risky a project is, however, the greater the chance that the firm would have conducted it anyway and the less effective the government subsidy is in stimulating innovation. The administrator therefore has an incentive to pick non-risky projects that the firm would have researched anyway in order to show off his acumen for spotting winners.

MacDonald (1986) shows in a case study of the Australian grant system for encouraging R & D that the grants are given to exactly the same kinds of projects that firms research anyway. Thus, he argues, the program loses much of its value.

Indeed a number of countries have R & D subsidy programs that have as their stated goal to subsidize commercially viable projects with no reservation against projects that firms would have conducted anyhow.

From the firm's point of view subsidies are generally welcome as extra income. There are a few reservations however. A great dependence on subsidies may weaken a firm's competitive edge. Many managers also claim that subsidy policies are not salient for their decision making, and that frequently they reduce the efficiency of projects due to bureaucratic constraints and delays (Rubenstein et al., 1977; Ettlie, 1983).[11]

Surveys

Differences in subsidies' effectiveness depend on how they are doled

---

[11] Roessner (1977) argues, and supports empirically, the notion that in some industries demand is so uncertain that even large subsidies will stimulate innovation much less than confirmed orders. This is shown in a study of firms dependent on local government for their orders.

out. Therefore it is important to correlate any findings about effectiveness with the type of subsidy program being studied.

Consider a few studies leading to opposite results. Mansfield (1984) reports a study of 41 federally funded energy R & D projects. He finds that firms would only have financed 20% of these projects themselves. Further each dollar increase of federal funding increased firms' R & D spending with 12 cents even though federally funded projects only appeared to add half as much to firms' productivity as firms' own spending on R & D.

Another study paints a much bleaker picture of government intervention. Gronhaug and Fredriksen (1984) examine the Norwegian innovation plan in existence since 1977. The plan includes grants covering 65% of project costs and low interest loans covering 85% of R & D costs which need not be paid back if the project fails. The projects were selected based on potential profitability, novelty of the project idea and the assumed R & D competence of the applying firm. The authors find that 78% of the projects would have been conducted anyhow, even though some of them on a reduced scale.

What can account for the difference in these studies? Upon closer inspection the two subsidy programs appear quite different. The Norwegian funds were granted to a variety of firms, each applying with their own research ideas. The government administrators in turn evaluated the projects in terms of commercial viability, without posting any own technological goals in the field.

The American energy support is different. Here the government came with a bag of own ideas, or developed ideas together with firms, in addition to supporting ideas originating in firms. Further the focus was less on narrow commercial viability and more on other goals such as developing techniques that could become viable in case of an energy shortage.

In interpreting either of these results one must remember that even if the part of unnecessarily subsidized projects represents a smaller fraction of total subsidies this can still place an intolerable burden on the efficiency of subsidies. A simple example can demonstrate this point. Suppose the firm increases the amount of its research by 50% of the value of the subsidy. Suppose further that additional research has a social return of 20%, remembering that these are marginal projects that the firm did not conduct without the subsidy. Then the social value of granting the subsidy is only 10% of the amount of subsidy. This may well be less than the social cost of raising the amount of the subsidy via taxes.[12]

---

[12] Hansson (1984) estimates the social costs of extracting taxes in Sweden at between 20% and 700% of the funds raised.

Econometric methods

The third method of examining the efficiency of subsidies are econometric studies comparing the extent of R & D in subsidized and unsubsidized firms. One problem of these studies is that it is difficult to infer a direction of causality. For example Scott (1984) finds, using U.S. data, that firms within each line of business firms that receive more government financing also conduct more own research. The problem here is that the government subsidies may not "crowd in" private research as inferred by the study. Rather firms with bright engineers may propose ideas that attract both firm funds and government aid. Lichtenberg (1984) attempts to correct for this by computing the correlation between the increments in non-subsidised and subsidized R & D to eliminate the time-independent industrial characteristics instead of using the levels of these variables. This is not a totally convincing technique because some changes in technological opportunities may favor increases in private as well as government R & D. In any case, this study finds that private R & D decreases when government subsidies are larger. A similar result is achieved by Carmichael (1981) and Levy and Terleckyi (1983). The latter conclude that government contracts and university research stimulate private R & D while subsidies seemed to reduce private research expenditure.

Using similar methods Holemans and Sleuwaegen (1988) find that government grants increased firms' R & D spending by 30–40% of the value of the grant.

All in all the econometric studies must be viewed with some caution both because of the variation in their results and because it remains uncertain whether they are picking up more than mere correlations.

4.3 The Swedish survey study

The study we summarize in this section is based on interviews with R & D managers in Swedish firms. This study is described thoroughly in chapter 9. Survey studies in general fall into two groups. One approach has been to query respondents about their general judgements concerning a policy. The other is to focus on specific decisions that the respondents have made and ask how they would have been changed in the presence of various policies. In this study we do both. This provides a control of the extent to which respondents merely draw on their general judgements when they reconsider specific decisions. The specific decisions in turn permit a quantitative estimate which is necessary for a judgement of whether subsidies are socially worthwhile.

The subsidy instruments that were compared are listed below.

Table 4.1 Subsidy instruments compared in the Swedish study

General subsidies
   1. Tax deduction for R & D expenses
   2. Grant toward costs of R & D personnel

Selective non-self-financing subsidies
   3. Project grants
   4. Project loans at low interest rates
   5. Conditional loans that are repaid only if R & D is succesful

Selective self-financing subsidies
   6. Fee-based loan guarantees
   7. Royalty grants. Royalty to the state is based on sales of the invention toward which the grant was applied.
   8. Stock option grants. In return for an R & D grant the state receives a stock option that can be exercised if the stock value rises significantly. For large firms the stock option refers to separate venture companies set up around the respective R & D project.

In the first part of the study R & D managers were asked which of these subsidy instruments they believed to be most effective in terms of generating the most R & D at the lowest cost to the public purse. The detailed results are shown in chapter 9. We do not reproduce them here since they by and large confirm the results of the quantitative study.

For the quantitative study R & D managers were asked to select a number of specific projects that were representative for the firm's overall R & D programme. Some of these were projects that firms had considered but decided not to conduct. Managers were then asked how the firms decisions to invest in certain projects probably would have been affected by the different subsidy instruments. The subsidy instruments were specified in exact economic terms so that managers knew how large a subsidy they could expect for each instrument.

Based on managers' responses it is simple to calculate the additional R & D that each subsidy instrument would induce if all who desired a subsidy received it. This procedure is appropriate for the general subsidies but not for the selective subsidies since the latter are granted only to those that, in the eyes of the subsidizing agency, should be subsidized.

The question is how well the subsidizing agency can select chaff from wheat. In the study we allow for three different levels of ability on the part of the subsidizing agency. It can either select perfectly only those projects that from a social viewpoint ought to be subsidized; or it selects randomly; finally it can select at a medium level of accuracy, selecting every second project perfectly and the rest randomly. In chapter 9 the results are shown for all three assumptions. Here we reproduce the results only for the medium level of accuracy. Table 4.2 shows how much additional R & D is generated by each subsidy instrument. This is shown for large and small firms seperately.

Table 4.2 Ratio of R & D generated by the subsidy to the present value of the subsidy

	Large firms	Small firms
1. Tax incentive	0.19	0.08
2. Grant to R & D personnel	0.16	0.07
3. Project grants	0.41	0.52
4. Project loans	0.40	0.59
5. Conditional loans	0.47	0.64
6. Fee-based loan guarantees	0.48	0.47
7. Royalty grants	0.56	0.74
8. Stock option	0.72	0.92

As table 4.2 shows the general subsidies do not generate much additional R & D. For example the tax credit generates new R & D only for 19% of the value of the subsidy. For small firms the effect is even smaller.

Apparently the most effective instruments are some of the self-financing subsidies. For example stock option grants induce R & D for nearly the entire value of the subsidy. The self-financing subsidies are also fairly insensitive to poor information on part of the subsidizing agency. The reason is that, given the conditions attached to the royalty grants and stock-option grants, firms are not interested in receiving the subsidy for many of the projects that they would have conducted anyhow. This "self-selection" means that it is less important for the subsidizing agency to make the right choice of projects.

4.4 An experiment on project selection

In this section we summarize an experiment concerning the way in which administrators of subsidy programmes select projects for subsidization. The experiment is described in detail in chapter 11. The two central questions that the experiment is designed to illuminate are how closely the judgements of different administrators match each other and whether a calculation of projects' expected social values aids project selection.

The central conclusion of this experiment is that project evaluators may often have a poor understanding of the factors that determine the social value of technology subsidies. This suggests that a considerable pedagogical effort may be necessary to improve the quality of project selection. Quantitative estimates of the value of granting subsidies may fill such a pedagogical effort regardless of whether they are of much use to a competent subsidy administrator in daily work.

The experiment was triggered by the decision of the Swedish Department of Energy to commission a computerized cost-benefit model (CCB) as a project selection aid in a subsidy programme to promote electricity saving technology. When the CCB was introduced it was possible to compare administrator's project decisions with and without the help of the CCB.

The experiment was performed using six actual projects. In each case a company or organisation had applied for a subsidy for projects that aimed to develop or demonstrate some electricity-saving technology. If succesful these technologies had the potential for wide imitation. The decision making problem for subsidy aministrators consisted of avoiding two mistakes. First, subsidies should not be given to projects that the applicants would have conducted anyhow. Second, subsidies should not be granted if the social value of demonstrating a technology is smaller than the required subsidy. Subsidies should thus be granted to projects that are too risky or have a negative expected value to the applicant but that still have a large social value. An important aspect of these projects is that they can have a positive expected social value even though the expected private value is negative. The reason is that if the technology turns out to be successful then it will be imitated by many, generating a large social value in the process.

The CCB is described in detail in chapter 10. Like all cost-benefit models the CCB weighs expected project costs against expected benefits. The unusual features of the CCB are first, that it explicitly allows for uncertainty in the input variables and, second, that it uses this uncertainty to estimate the expected social value of the projects.

For each variable a user enters a minimum, medium, and maximum value reflecting the uncertainty perceived by the decision maker. This is done for all costs of the project using the new technology and for all costs of the best alternative to this technology.

The CCB then calculates the probability distribution of the difference in costs between the project and the alternative. From this one gets the probability distribution of the net present value of the project. The CCB then calculates a diffusion curve for the new technology based on empirical experience with similar technologies. The diffusion curve is a function of the project's net present value. If the project incurred a loss no diffusion occurs; otherwise the speed of diffusion is an increasing function of net present value. From the distribution of net present values one has thus derived a probability distribution of diffusion paths. Finally the CCB calculates the expected social value from the distribution of diffusion paths.

Five subjects with considerable experience of project evaluation and a good knowledge of the technologies involved were asked to estimate each of the six projects' private and social values. Then they were asked to enter their estimates of input costs into the CCB. In total this resulted in ten estimates of each project's private and social value.

The precise results are shown in chapter 11. We do not reproduce any tables here but let a qualitative summary suffice. There seems to be a clear pattern in the results. The CCB and project evaluators come to rather similar conclusions about the private expected value of the projects. Their judgement about the projects' social value however diverges considerably. The latter effect is especially pronounced for projects that are expected to have negative private values. Compared to the CCB, the project evaluators exhibit a considerable bias to believe that the social value was negative if the private value was negative.

There are two possible interpretations of these results. One is that the CCB misses some aspect of the social value that project evaluators capture. This interpretation is contradicted by the fact that project evaluators themselves do not come to similar conclusions about the social value.

The other interpretation is that project evaluators do not have a good understanding of the factors that determine the social value of technology subsidies. This interpretation is supported by post-experimental interviews. The interviews showed that the subjects often did not have a good understanding of what determines the social value of the subsidies. Often the subjects accepted the CCB estimates of social value once the reasoning was explained. This would imply that the CCB fulfills an important pedagogical purpose regardless of whether it is practical in everyday use.

4.5 Other innovation policy issues

In this section some empirical evidence on innovation policy issues is summarized that are somewhat peripheral to the main course of this book. Primarily we examine the effectiveness of patents and of government-financed research institutes. These topics are included here because they are often seen as the main alternatives to subsidizing R & D in firms. It is sometimes believed that prolonged patents can serve as a substitute for government subsidies. Similarly government R & D may serve as a substitute.

Patents

The rationale behind the institution of a patent system rests on the recognition that technological knowledge has certain attributes of a public good. From this perspective, knowledge, once created is believed to be freely appropriable by others and the "free-rider problem" thus limits the incentive to create new knowledge. By conferring property rights that restrict temporarily the wide use of new knowledge, the patent system is supposed to create the incentive to engage in inventive activity and to undertake the costly investment typically required to reduce an invention to practice.

In fact, empirical research, especially that of Taylor and Silberston (1973), Mansfield et al. (1981), and Levin (1986) has made it clear that patents rarely succeed well in conferring appropriability. Many patents can be "invented around". Others provide little protection because they would fail to survive a legal challenge to their validity. Still others are unenforceable because it is difficult to prove infringement. Griliches and Pakes (1987) argue that patenting cannot on the whole be very important for innovation since the value of patent rights generated each year is only in the order of 10–15% of national expenditure on R & D.[13]

Levin et al. (1987) shows in a survey of 650 R & D executives that patents were viewed by R & D executives as an effective instrument for protecting the competitive advantages of new technology in most chemical industries, including the drug industry, but patents were judged to be less effective in most other industries.

Mansfield et al. (1981) finds that in manufacturing industry the existence of a patent was thought to raise imitation costs by only 6% on average. At the same time Mansfield concludes that imitation costs can sometimes be substantial in the absence of any legal protection. Therefore the patent does not even insure eventual diffusion of the technology after the patent has expired. In a further study of 100 U.S. firms Mansfield (1984) finds that information concerning development decisions is generally in the hands of competitors within 12 to 18 months, on the average. Information concerning the detailed nature and operation of a new product or process leaks out within about a year.

In spite of the fact that firms often judge other factors than patents to be more effective in safeguarding the value of an invention they are in general favorably disposed toward the patent system. This is especially true in Japan where patents both are ranked as having a higher value in safeguarding technology and are used more often for defensive purposes to prevent other firms from inventing around inventions (Bertin & Wyatt, 1988).

If patents often do not work well why then do firms use them?

---

[13] Interestingly Griliches and Pakes find that while the number of patents declined in many countries during the seventies the average value of patents increased. Their "quality-adjusted" index of patents actually rose during the seventies.

Further study is needed, but one possible answer is that patents are useful for purposes other than establishing property rights. Patents may be used to measure the performance of R & D employees, to gain strategic advantage in interfirm negotiations or litigations, or to obtain access to foreign markets where licensing to a host-country firm is a condition of entry.

In any case these findings imply that the patent system has modest effects on firms' incentive to research. Thus lengthening patent lives, or trying to fine-tune them to indiviudal industries is a fairly uninteresting policy issue. The most that can be said perhaps is that given the relative ineffectiveness of the patent system, the bureaucratic costs of filing patents should be kept to a minimum. One should not put down a lot of effort on a losing team.

Government research institutes

Government research institutes serve two purposes: To conduct R & D that is socially valuable but that the firm does not conduct on its own and to help diffuse know-how. The argument is that research results attained in a research institute rather than in a firm can be spread to other firms instead of remaining secret. In particular small or medium sized firms, too small to research on their own, can order research from the institutes that also benefits other firms. A number of surveys have indeed found that industrial firms identify research institutes as the most frequently consulted source of extra-mural scientific and technical information (see Pavitt and Walker, 1976; Rothwell and Townsend, 1973).

Two main problems, reflected in the applied literature, afflict the government research institutes. First, researchers working in the government institutes may research the wrong projects. This may be because they have wrong incentives. For example they may be more interested in publishing articles than in designing new widgets. Or they may lack the firms' knowledge of what inventions are commercially viable.

Second, the kind of knowledge needed by industry to produce a new gadget may not be easily transferred. While the "know-why" of science is easily disseminated, the know-how of technology is locked up in individual employees' experience. Even when the knowledge is easily transferred the research institute employees may not have the right incentives to distribute this information and localize all potential users.

Much of the applied literature on government research institutes consists of case studies. Some of the more systematic attempts to evaluate government research institutes (e.g. Toren & Galai, 1978) typically find that the institutes are most useful for medium-sized companies in low-technology branches. Small firms often lack the resources even to engage in contracted research at an institute. Large firms on the other hand do their own research. In high-technology branches firms are eager to keep research secret, so they do not like to involve outsiders.[14]

The problem of information flows is examined in studies by Allen et al. (1983) and Leonard-Barton (1984). Allen et al., in a multi-country comparison show that technology flows principally through informal channels within industries. Very little of the total information flow was obtained from the formal mechanisms or institutions, such as research institutes, normally considered central to the technology transfer process. These studies focussed on high-technology firms which may explain the contradiction with the results referred to above.

The U.K. Department of Trade and Industry offers a number of awareness and consultancy schemes, the most important of which is the Manufacturing advisory Service (MAS). Between 1977 and 1982 1043 MAS projects were completed at a total cost of 4.7 million pounds. On the basis of informed feedback it achieved a benefit-to-cost ratio of 12:1 (Rothwell, 1985).

Clearly, there is a dearth of empirical studies concerning research institutes. Since research institutes are somewhat peripheral to this book we refrain from discussing the large literature on research institutes that is not grounded on proper empirical evaluation.

---

[14] A typical case study of research institutes is the following. In 1980, Norway called in a British team from the Science Policy Research Unit (SPRU) at Sussex University to do an independent audit of its government-backed research institutes. Although SPRU politely concluded that Norway runs such government labs better than most countries do, its report makes depressing reading. It details many projects which the institutes developed with industrial applications in mind, but which industry either did not take up at all or took up unsuccessfully. The report lists successes too. Several arose when research-institute employees left to join, or found, the company exploiting the innovation (one company founded was the computer firm Norsk Data).

Part three: Reforming innovation policy

The following four chapters formulate a catalogue of policy prescriptions. Readers who are content with a summary of the policy prescriptions may turn directly to chapter 8. The other three chapters, 5 to 7, assess the arguments that lead up to the policy prescriptions.

Some of the policy arguments are firmly based on empirical and theoretical results. Others find their roots in the author's subjective experience. The latter is particularly true of chapter 5 in which an attempt is made to describe, in a stylized way, the political forces that shape innovation policy. Many readers will undoubtedly question the objectivity of such a description, especially since it is meant to be a valid description of the situation in many countries. They are right to be sceptical. Yet most will agree that any reform of innovation policy must begin by asking whether the process by which policy is formulated and executed is sound. Many innovation policy experts feel it is not and chapter 5 is an attempt to sum up some of the criticism.

Chapters 6 and 7 are supported much more firmly by evidence. Chapter 6 suggests measurement techniques that can be used on a regular basis to assess the efficiency of subsidy policies. In particular the use of classical experiments is evaluated. In addition principles are discussed for assessing the social benefit of individual projects that subsidies give rise to.

Chapter 7 suggests principles that a subsidy instrument should adhere to in order to be efficient. These principles are derived from the theoretical analysis in chapter 12. In a comparison of practically used subsidy instruments it is shown that those that adhere best to these theoretical principles also seem to be those that perform best in empirical investigation.

5 The politics of innovation policy

5.1 Introduction

Economic theorists discussing problems of public decision making have focussed almost exclusively on the role of incentives. The absence of the profit motive in the public sector, they claim, inhibits effort and initiative. This view has frequently met with little understanding from public employees who retort that the work effort required of them is often at least as high as in the private sector.

In the part of the public sector concerned with innovation policy both views may be partially correct. This chapter aims to show that the real dilemma of innovation policy has been the failure to collect information about the efficiency of innovation policy. The lack of incentives to test and evaluate policies has probably been a hundred times worse than any deficiency of incentives for individual work effort. In that sense there is an incentive problem but not of the kind most commonly conjured up by theorists.

In this chapter we analyse the process by which innovation policy is formulated and suggest a few measures for reforming the process. In particular the establishment of an independent body is recommended that evaluates innovation policies using scientifically accepted methods.

5.2 The formation of innovation policy

The formulation of innovation policy is a game played by four categories of players: The government, government agencies executing the policy, firms, and finally, independent experts, usually academic professionals.

Usually a policy change only occurs when one of these groups has vented considerable criticism of a policy and has managed to gather broad support for action. Once an issue has attracted a threshold level of interest the demand for action suddenly becomes intense. At that point policy analyses and proposals are often produced rapidly. Their content will depend much on the political mood. Such a hurried decision will often not be a wise one. Yet once a policy change has been instituted it usually remains in place for quite some time regardless of how well or poorly it works. It is then impossible to gather the political impetus required to institute a change a short time after a previous change.

Ideally the process of policy formation should follow quite a different pattern. A continuous reevaluation of policies should indicate policy problems before they become acute. New policies should be tested on a small scale and evaluated using objective methods.

I do not wish to paint too murky a picture of the policy process here. Some experimentation does occur, and occasionally some excellent policy analysis is performed. However only in rare cases are all elements of the ideal development present: A sound policy analysis leading to a well designed policy experiment that includes a control group and scientifically acceptable measurement methods, followed finally by an appropriate follow-up of the results.

This ideal is rarely adhered to in any area of public policy. Innovation policy is, even more than most other policy areas, one in which there is no natural opposition to greater public spending. None of the four groups of players have a vested interest in reducing the extent of innovation policy.

As a result a typical round in this game takes the following form. The government is under pressure to show some initiative in raising industry's ailing productivity growth. It has a general ambition to be restrictive with public funds, but compared to the major public spending programmes an R & D subsidy programme is a relatively cheap way of showing initiative. In addition it is unlikely to encounter opposition. The government agency in charge of R & D subsidies is only happy to oblige. Firms accept the hand-out without protests regardless of how ineffective they believe the subsidies to be. Independent experts on innovation policy do not protest too loudly either since they are, to large extent, dependent on consultancy contracts from the government or the subsidizing agency. After all, what use is innovation policy expertise if innovation policy were abolished?

In the following we examine the players' incentives more closely. Begin with the role of the government. The government rarely has much expertise of its own. Rather it is forced to rely on the expertise of government agencies, firms, and outside experts. This lack of expertise, and lack of concrete proof of instances of innovation policy failure, often makes it difficult for the government to withstand pressure for greater spending. Even when the government has chosen an informed innovation policy the executing agencies usually enjoy considerable leeway in applying it. Thus the government suffers both from not knowing much about what policies work and from not being able to control in what spirit policies are applied.

The government's basic mode of operation is to observe problems and allocate money to those chosen to solve the problem. The so called "public choice" school of economics expounds a host of arguments for why governments have mistaken incentives leading to inefficient decisions. Without discussing these in detail there seem to be two main ways in which wrong incentives distort innovation policy. First, governments are often concerned with being seen publicly as doing something for innovation rather than with the question of whether what they do yields results. This seems to be a recurring complaint voiced by the case studies reviewed in chapter 4.[15] Frequently, it is argued there, money is poured into showcase projects with little concern for long-term viability.

Second, governments may be unwilling to rule against the vested interests of vocal special interests. As a result governments do not usually make decisions that are strongly opposed by the policy executing agencies or by firms.

After this extremely rudimentary description of the government's role in innovation policy consider the part played by the other categories of agents.

Government agencies

The public bodies executing innovation policies are often heavily staffed with engineers. That in itself means that questions of economic efficiency often do not receive high priority. When they do, it is usually private rather than social economic efficiency that receives attention. For example, innovation policy administrators often interpret their mission to consist of supporting technologies that they expect can be profitable to firms. What makes an R & D project profitable to society rather than the firm is often not well understood and is rarely incorporated into the design and targeting of innovation policy.

More important for the way these bodies work, however, is their near monopoly over the three functions of executing, evaluating, and suggesting reforms of innovation policies.

In policy execution employees rarely have time or resources to conduct effective evaluation. They are rewarded for smooth administration which sometimes leads to crass goals. The main ambition is surprisingly often to ensure that the budgeted funds are dispersed.

Usually officials, seasoned by years of executing policies, eventually climb the ranks and come into the position to influence decisions on innovation policy. The normal style of management in such an agency, and probably the most effective as far as day to day administration is concerned, is to try to create some consensus around the organisation of work. This way of approaching a problem is then applied even to policy evaluation. As a result evaluations are usually conducted haphazardly based on interviews with the involved parties. The purpose of policy change becomes to create a compromise that pleases all involved parties; alas no party has a direct interest in promoting what is best for society.

---

[15] Some good examples are given in Roessner (1984). Other case studies with similar implications are reviewed in chapter 4.

Firms

Firms usually have an ambivalent view of subsidies. On the one hand they loathe government involvement. On the other hand a subsidy helps profits and competitiveness. Firms usually resolve this dilemma by lobbying for unconditional aid such as tax relief in their own industry – and none in other industries. It is interesting to note, however, that industrial federations or lobbying groups in several countries actually oppose R & D subsidies to firms. Instead they often propound more technology-oriented public procurement.

Outside experts

Economists have displayed a considerable lack of interest for detailed studies of efficiency of innovation policies and subsidies in general. In part this may be explained with the intellectual roots of economics as a study of the static efficiency of markets. In the pursuit of this intellectual tradition economists have rewarded each other for generalizing results as much as possible. One consequence of this tradition is that many conclusions are so general that they are of no use for detailed policy design.

More important however is the availability of data. Modern empirical economists flock in droves to macro- and labor market issues simply because data are available in those areas. In contrast data are intrinsically hard to get in the area of innovation. Main stream economists have resisted unfamiliar methods of data collection such as surveys. In addition government agencies have often been less than helpful in securing such information.

Finally, innovation policy experts have been shaped by the fact that subsidizing agencies are their main clients. In part, a selection process is at work. Subsidizing agencies pick experts who do not question basic premises, and these experts who receive agency consultancy contracts eventually build a name as prominent innovation policy experts. In addition to this selection process most innovation policy experts find that there is no market for studies that could call the need for subsidies into question. The main markets for consultancy contracts concern what technologies should be subsidized and which type of firms should be aided.

This brief description of the political process behind innovation policies summarizes the experience of the author and many innovation policy experts that the author has talked to. Undoubtedly many readers would add something, place the emphasis differently or even disagree entirely. They are in their full right to do so since we have presented no provable facts. Nevertheless even a substantially altered description of the policy process must somehow explain the mechanisms that leave the innovation policy apparatus so uninterested in ascertaining how efficient innovation policies actually are.

5.3 An independent policy evaluation

A greater mind may one day invent a way of providing the government and government agencies with incentives to choose unswervingly what is in the best social interest. Until then more modest aims must suffice. Clearly a large stride toward more efficient innovation policies is achieved by providing better information about what methods work best.

Our description of the policy suggests that such information will not be forthcoming from any of the interested parties. What is needed instead is an evaluation agency that on the one hand can maintain a scientifically acceptable standard and be free from the pressures that the policy executing agencies are subject to. On the other hand this agency must have the clout to move the executing agencies to perform policy experiments, to document their operations in a way that makes efficiency comparisons possible, and to gain access to the agencies' confidential data.

These requirements suggest a government agency staffed by a mixture of innovation policy practitioners and academics. This body should recruit its employees largely from outside the circle of established administrators of innovation policy. Further the methods of the evaluation agency should be constantly reviewed by academic economists and others.

The modus operandi for this body should be to design and evaluate policy experiments and to refine and apply the social cost-benefit methods that are relevant here. These are described in more detail in the next chapter.

In contrast to general government evaluation agencies such as the General Accounting Office, this evaluation agency would engage in a long-term build-up of data bases and evaluation methods.

Apart from pure evaluations the evaluations agency serves to spread the insights it generates. The pedagogical effort required to convert the empirical insights into routine procedure in the subsidizing agencies is probably a more formidable task than the evaluations themselves. Thus it can provide the legal and financial expertise that the subsidizing agencies may need occasionally to apply specific subsidy forms. In particular those subsidy forms that entail some payback provision require a sound legal backing.

The evaluation agency could also perform a number of coordinating tasks that today are neglected in most countries. The most important of these is to provide a common database for all the different subsidizing agencies. In this database a subsidizing agency could check instantly whether an applicant has received subsidies from others, who the main competitors are to a specific technology, who the experts are in various technological fields that are capable of judging the merits of a project and so forth. The evaluating agency needs such a database anyway to prepare summary statistics and it would be a simple matter to give subsidizing agencies access.

A further coordinating role that the evaluation agency could serve is as a common bank. Today the prevailing system is that evaluating agencies receive budgeted funds that they then are eager to spend. There is little opportunity for reallocation during a budget year if it turns out that one subsidizing agency receives much more promising subsidy application than another agency. If the evaluation agency served as a common bank this problem could be ameliorated. Subsidizing agencies could then request funds at the rate at which they receive interesting applications. The evaluation agency could continuously signal how good applicant projects have to be to receive further funding. Thus one could achieve a rough reallocation between subsidizing agencies.

At this point it may appear that a major issue has been circumvented: Who decides how much each subsidizing agency can disburse? As described above the evaluation agency has a measuring and coordinating role but it is not a central decision maker. Rather the government budgets funds to each subsidizing agency as is common practice today. The new elements are first that the government makes its decision under consideration of the evaluations performed by the evaluation agency and, second, that it provides a possibility for subsidizing agencies to release funds that the evaluation agency then can reallocate to other agencies.

The crucial aspect of this system is that subsidizing agencies must prefer to release funds rather than being criticized by the evaluation agency for granting subsidies that are not socially worthwhile. The government can achieve the proper incentives by threatening to close down subsidy agencies that consistently allocate subsidies poorly. In contrast an agency that often releases budgeted funds may receive a cut back in budgeted funds but may survive with its name in good repute.

The threat of closing down subsidizing agencies may have to be executed occasionally. In turn one can allow new ones to begin operating.

One may even ponder the possibility that the evaluation agency loses its sting. Undoubtedly even this agency may mutate into a useless bureaucratic vegetable. Experience with similar bodies suggests that this is a real threat. Since this agency does not itself make administrative decisions, its role being to gather and broker information, there is no reason why it should not be run privately on the basis of limited contract periods. Alternatively evaluations can be conducted competitively by private consultants or academics that are hired by the evaluation agency. This introduces at least some competition into the process.

6 Performance measurement

6.1 Introduction

The purpose of this chapter is to make the previous chapter's demand for better policy evaluation concrete. The first part discusses methods for comparing one subsidy policy to another, in particular through the use of classical experiments. The second part suggests ways of estimating the effect of a given subsidy on a specific project.

Much of what is sold as policy evaluation in the subsidy business is really nothing more than an intelligent summary of interviews with a few subsidy administrators and recipients. For obvious reasons these evaluations easily fall prey to a lopsided concentration on administrative issues. Also they rarely question whether a subsidy is socially worthwhile. True policy evaluations should concentrate on measures that are more objective than gathering opinions of partial insiders.

Ideally the preparatory stages of policy design should include some kind of "evaluability assessment". This means that policies should be designed with an eye to how easily and objectively they can be evaluated. Preferably an objective evaluation should be organized as a classical experiment.

6.2 Experimental policy evaluation

There exist essentially three scientifically acceptable methods to evaluate the efficiency of subsidy instruments: Surveys, econometric analyses, and classical experiments. Other methods, such as case studies or collecting a few informed opinions, are helpful in some instances, but cannot be considered as weighty evidence when judging the efficiency of subsidies. The latter methods are useful in situations one expects the case study to be representative of the population one analyzes. Subsidies however seem to have very different effects on different firms and projects so it is important to base conclusions on average results from statistically significant samples.[16]

As described in chapter 4 most attempts to estimate the effectiveness of subsidies have used the first and second method, surveys and econometric analyses. Classical experiments are rare. In a sense this is surprising since such experiments are usually considered to be the "ideal" scientific method. In related areas such as estimating the efficiency of subsidies in the labor market a number of such experiments have been performed.

---

[16] Some aspects of subsidy policies may be much more regular and therefore more amenable to analysis by case study. For example, a number of interesting case studies have studied the political process that led to a subsidy being granted.

Some experience with innovation policy experiments has been garnered by two ambitious experimental programmes in the United States. One programme was being conducted by the National Science Foundation (Experimental Research and Development Incentives Program) while the other was administered by the National Bureau of Standards (Experimental Technology Incentives Program). Even though they produced a number of excellent policy experiments their overall success was rather limited. Often political pressure for early success distorted programme objectives and encouraged policy demonstrations under conditions favoring "successful" outcomes, rather than true policy experiments (Robbins and Milliken, 1977).

The ETIP (Experimental Technology Incentives Program) was terminated after 10 years of operation after internal management problems and waning political backing (Tassey, 1985). These experiences emphasize the importance of having an evaluation agency that is well insulated from direct political pressure.

In the remainder of this section we discuss in some detail the practical considerations that must be entertained when designing an experiment. This discussion draws on the theory of experimental methods, experiences from experiments in the labor market, and the author's experience with a variety of experiments including the one reported in chapter 12 and summarized in section 4.4.

The simplest, and most persuasive, classical experiments consist of two groups that are equal in all respects except in terms of one controlled experimental variable. An example would be to announce a subsidy policy and divide applicants randomly into two groups. One of these groups receives the subsidy and the other does not. A year later one conducts a survey among firms and determines the extent to which they conducted the R & D projects that they wanted the subsidy for. This ideal arrangement is not always easily reproduced in practice. There are a number of common problems and a number of ways to deal with them.

Measuring effects of a subsidy

The first problem is to decide what variable to use to evaluate the effects of the policy on the experimental group. In the case where one is comparing two groups of firms, one of which received subsidies there are usually two sets of variables that can be compared. The first set concerns firms' long-run performance, such as growth, profit, or stock appreciation. The second set concerns firms' short run R & D decisions such as the amount of R & D conducted, the number of researchers employed, and the type of research projects accepted. The second set of variables is in general related more closely to the effects of an innovation policy and will probably give more reliable results. The problem is that these variables by themselves do not give a good picture of the social value of the policy. For example, if one finds that the experimental group, the one subjected to a policy, researches more, one has no conclusive answer to question whether that extra research is worth the social effort. This problem is of course common to all empirical methods and it is not as serious as it appears at first. Even when the social value of a subsidy cannot be ascertained it is possible to compare different subsidy instruments in terms of the measured variables.

One possibility is to use the cost-benefit methods described in the next section to gauge the social value of the extra research generated.

Representativeness

The experimental group should be representative of the population to which the policy is applied. This requirement can be problematic because it sometimes conflicts with the just administration of policies. If a subsidy programme is normally sufficiently endowed to support all applicants then conducting an experiment requires that some randomly chosen firms are not given subsidies in order to serve as a control group.

A further problem is that the mere knowledge that an experiment is being conducted may affect the composition of the sample of firms that apply. It may of course not always be necessary to advertise the fact that an experiment is being conducted.

Hawthorne effects

Hawthorne effects are changes in peoples' behaviour that occur merely as a result of the fact that an experiment is being conducted. The term "hawthorne effects" stems from experiments on the impact of the work environment on productivity. It was found that employees found any change in their work environment stimulating.

In connection with innovation policy experiments there is a risk that subsidy administrators behave differently when they know that the experiment is being conducted. They may select or check subsidy applications more carefully; or they may be reluctant to thoroughly learn the administrative techniques that the experiment requires. A particularly worrying possibility is that administrators have a personal stake in the outcome of the experiment. They may themselves have been involved in developing the policies that are being tested; or they may be worried about the consequences of a new policy, if shown to be succesful, for their own job situation.

Hawthorne effects can be minimized if careful consideration is given to the details of an experiment and who conducts it. Further one can sometimes introduce "blind" selections that filter out hawthorne effects.

Indirect effects

Some policies raise the level of innovation for some at the expense of innovation for others. The most prevalent example of such an effect is that firms that receive public support for R & D employ researchers that are not available for other firms. This effect is more worrisome when large amounts of public funds are poured into narrow areas where the supply of well educated researchers is very inelastic. In studies where one is comparing the effectiveness of different policies in generating additional R & D it may be justified to ignore this effect, assuming that it affects all the policies roughly equally.

Indirect effects may well be positive. For example the R & D generated by public support in one firm may stimulate R & D in other firms.

For the purpose of conducting experiments one is usually forced to ignore the wider and more diffuse indirect effects on unspecified other firms. One should however be careful to avoid indirect effects that arise merely due to the administration of policies. For example, the experimental group should not receive preferential service from the subsidizing agency beyond what the experimental treatment requires.

Costs

Experiments are often costly because they involve developing and administrating a new policy. In most countries however such policy renewal already occurs regularly. What is lacking however is a control group and a careful measurement of the effects of a new policy. This means that there may be a very low price tag to merely organizing the policy changes that already occur as proper experiments.

The main obstacle in some cases is the selection of a control group. It requires a discrimination between firms with equal rights to the benefits of an innovation policy. Sometimes this can be handled smoothly by applying a different policy to the control group, so that one is comparing two policies rather than one policy compared to no policy.

Quasi-experiments

Rather than conducting a true classical experiment one can sometimes conduct so called quasi-experiments. This means that one examines an already existing policy and finds a control group that is similar to the group subjected to the policy. For example if the policy was confined to a certain region one might find similar firms in a different region. For an interesting example of such an experiment see Bohm and Lind (1988).

6.3 Project evaluation

A wide range of quantitative techniques have been proposed for judging the merits of R & D projects. In general these have not met with much appreciation from firms let alone subsidizing agencies. Reviews of these methods can be found in a report by the Office of Technology Management (1986). Instead peer review methods are usually recommended.

Many subsidizing agencies today let their administrators judge the technical merits of projects they subsidize. Often this produces worse results than a peer review process because the administrators usually do not have the necessary specific knowledge. Peer review does have its problems however. It is highly questionable whether it is desirable or possible to use it as the only project selection method. The first problem with peer review is that it may not be very responsive to the needs of new fields of research. A number of studies conclude that peer review allocated the greatest funding to a field of research when the most exciting discoveries were already made and the field was declining in importance.

The other problem of the peer review process is that it violates the secrecy of research projects and may therefore not be applicable in many instances where firms are subsidized to develop novel ideas.

In this section the use of some quantitative methods is suggested. These methods are expounded in detail in chapter 10. Here we supply only a brief description.

Numerous quantitative techniques for R & D project selection have been suggested (see e.g. Sounder & Mandakovic, 1986). None of these has become very popular. For the purpose of selecting projects for subsidization the interesting concept to quantify is the expected social value of subsidizing. In many other areas the social value of public investments is estimated with social cost-benefit analyses.

Social cost-benefit analysis involves a set of techniques for estimating social rather than private values. For some applications it has its obvious use. For judging R & D projects however its usefulness is strongly questioned. The main criticism are the following:

1. The benefits and costs of R & D projects are extremely uncertain. As a result the point estimates of costs and benefits are often no more than wild guesses.

2. Cost benefit analyses can be cumbersome to produce. This effort may not be worthwhile if it only produces wild guesses.

These two criticisms are probably valid as far as conventional cost benefit techniques are concerned. They ignore two arguments however. The first is that cost benefit analyses are excellent educational devices for teaching subsidy administrators what factors make subsidies socially valuable. The second is that newer cost-benefit techniques have been devised that are less cumbersome to use and that account for uncertainty about costs and benefits.

Our experiment, reported in section 4.3 and chapter 11, indicated that subsidy administrators sometimes do not have good grasp of what determines social value. This means that they can continue indefinitely granting subsidies to the "wrong" projects, those where granting the subsidy generates little social value.

The experiment compared the judgements of subsidy administrators with the judgement of a computerized cost benefit system (CCB). After the experiment subsidy administrators were taught the principles behind the CCB. According to their own statements that significantly enhanced their understanding of the concept of social value and, they said, would probably change the way they selected projects in the future.

This example shows that cost benefit techniques can serve as a type of "simulator," teaching practitioners what to aim for in practical work. Beyond that cost benefit techniques can be useful for actual project selection.

The CCB which is described in chapter 11 is an example of a cost benefit system that is totally computerized and standardized to fit a certain type of R & D project. It does not require more input information than a normal subsidy application anyway requires. Given this information the entire cost benefit analysis can be performed in less than half an hour. This means that the second criticism of cost benefit methods presented above is eliminated.

The CCB also handles grave uncertainty about the magnitude of costs and benefits. A user enters his uncertainty about input values. This is done by supplying three values: An expected value, a minimum and a maximum value. The CCB uses these to construct a probability function for each input variable. Based on these it calculates and shows how uncertain the private and social values are and what the expected value is given this uncertainty.

The final estimates of social value that the CCB produces therefore are values that incorporate the uncertainty about all variables. This eliminates even the first criticism.

As an administrative tool the CCB has even a third advantage: Since it records the estimates for each input variable it becomes immediately transparent to all on what basis the subsidy decision was made. This transparency is invaluable for attempts to judge the efficiency of subsidy programs.

In summary, modern cost-benefit methods can be easy to use and can incorporate uncertainty. They may help to assess the expected value of projects; but even when they do not lend greater certainty to the likely outcome of a project they are an invaluable pedagogical tool and make subsidy decisions transparent to others.

7 The custom design of subsidy systems

7.1 Introduction

Policy recommendations sometimes espouse the idea that rules governing innovation subsidies should be as uniform as possible. This cuts administrative costs, it is argued, and is appreciated by the recipients of subsidies. Typically such policy recommendations avoid the issue of how efficient subsidies are.

This chapter begins by describing principles that in theory make subsidies as efficient as possible. These principles in fact turn out to be fairly simple and to be roughly the same regardless of what the specific aim of the subsidy is. Ironically however, to uphold these principles in different instances of subsidy policy requires that the subsidy instruments and rules be allowed to vary considerably. The same rules can provide quite different incentives in different circumstances.

Section 7.2 discusses the theoretical principles. Section 7.3 shows how to apply these in practice and 7.4 discusses subsidy targeting and administrative rules. The theoretical principles are derived at greater length in chapter 12.

7.2 Theoretical principles

The theoretical aim is to devise a subsidy policy that will achieve an increase in R & D effort at a low cost to the public purse and without incurring undesirable side effects. This theoretical aim can be restated more loosely as follows. The subsidy policy should avoid the following three mistakes:

1. Subsidise projects that would have been conducted even in the absence of subsidies.
2. Subsidise projects that are not socially valuable.
3. Subsidy rules that do not give incentives for mismanagement of R & D.

If the subsidizing agency were perfectly informed about each project that firms apply with then these mistakes should easily be avoided. One would simply choose those projects that are socially most valuable and that firms would not have conducted of own accord. Further one could grant the subsidy conditional on the firm researching the project as promised. In sum, with perfect information the government simply purchases some R & D effort just as it would any other good.

In most cases, however, the subsidizing agency is poorly informed about the projects it subsidizes. The question is then whether subsidies can be designed in a way that helps the subsidizing agency to select the right projects, primarily by giving disincentives to apply with projects that should not be subsidized anyway.

The main insight about the design of efficient subsidy policies rests on the embarrassingly simple observation that the information a government agency has improves dramatically during the course of the project's execution. With the benefit of hindsight it is much easier to tell whether a project should have been subsidized or not. Based on this observation the idea is to attach a reward mechanism to subsidies that rewards firms after the project has been completed if it turns out that the project is one that should have been subsidized. Alternatively firms can be punished if the project should not have been subsidized. If the reward mechanism is made to function perfectly then firms will never apply with projects that should not have been subsidized.

Even though we have expressed this subsidy mechanism in terms of abstract rewards and punishments one should not be misled into thinking this to be some theoretical aberration. These terms turn out to have quite simple and commonsensical interpretations in practice. More will be said about those later.

Economists have in recent years been fascinated by so called "incentive compatible" policies, policies that give a policy target incentives to do exactly what the policy maker intends him to do even though the policy maker has poor information about at least some determinants of the target's behavior. In chapter 12 we analyze in some detail the design of incentive compatible subsidy policies. Two main conclusions emerge. The first is that a perfectly incentive compatible policy is impossible if one assumes that the subsidizing agency has no independent information about the projects it subsidizes before they are conducted but perfect information afterwards.

The second conclusion is that although a perfectly incentive compatible policy is impossible near substitutes can be found. One of these is the so-called "incentive subsidy".

To understand the theoretical principle better consider the following somewhat stylized assumptions. There is a set of projects that firms could research. Each of these has an expected private value. Firms will not conduct projects of own accord that have negative expected values. Also they will not conduct those that have positive expected private values but that are deemed too risky.

Each project also has a social expected value. We can assume the social expected value to be higher than the private expected value on the grounds described in chapter 3, for example because the invention can be used by other firms as well. Then the incentive subsidy should as often as possible give firms incentives to apply only with projects that have positive social values and either negative private values or that are too risky for firms.

Under the incentive subsidy scheme firms must apply prior to the commencement of a project. At that time firms may or may not receive an advance. The important thing is that the exact size of the subsidy is not determined until after the project has been completed.

The incentive subsidy contains a component that compensates the firm for a loss or taxes away a gain it makes on the project. In addition the firm is rewarded a fraction "a" of the social value. The subsidy g can then be written as follows, where R is the private return, and the tax of profit or compensation for loss corresponds to $-R$:

$$g = -R + aS$$

The term a S should be interpreted as merely some reward to doing what is in the best social interest. In many instances a rough approximation to the social value will be quite sufficient to induce the desired behaviour.

To see why the incentive subsidy works consider the firm's expected value of researching with a subsidy. We denote this expected value Rs, and then

$$Rs = E(R + g) = a S^e$$

Here S^e is the expected social value. This makes clear that the firm will not apply with any project that has a negative expected social value.

Since the firm is rewarded for maximizing the social value it also conducts the project efficiently, minimizing costs and maximizing the social value of the innovation.

When a project has positive private return then the firm usually looses by applying to the subsidy system because the private return will be taxed away. However, there is a special case, as mentioned above, where the incentive subsidy is not perfectly incentive compatible. The firm will lie about some projects it would have researched even without the subsidy, and will receive funding for them. If the firm is risk neutral this occurs for projects that have an unsubsidized expected return between 0 and a Se. As described in chapter 12 a can usually held rather low so that not many projects fall into this category. Note that the incentive subsidy also acts as an insurance. In some cases firms may subscribe to the subsidy merely for the insurance component. In that case the government can expected the subsidy scheme to go with a profit.

In chapter 12 the incentive subsidy is compared to other subsidy schemes using a simulation of hypothetical firms and research projects. Firms are assumed to decide whether to research or not under the different subsidy schemes in an attempt to maximize expected profits. Table 7.1 shows the increase in social value that a subsidy produces relative to the unsubsidized outcome. The table also shows what happens under different assumptions about how accurately the subsidizing agency is expected to ascertain the social value and the private profit of the completed project. The government is assumed to either to have perfect information, or make a random error, or make a systematic error.

Table 7.1 Percentage increase in social value over the non-subsidized outcome

	Perfect information	Random government error	Systematic government error
Incentive subsidy	26	16	9
Normal subsidy	19	–6	2

Clearly the incentive subsidy performs better than its competitors, especially when the subsidizing agency is prone to a large random error.

Turning from theory to practice the next section shows how to apply the rather abstract incentive subsidy in forms that can be readily administered.

7.3 Practical application of optimal subsidies

Taken literally the incentive subsidy described in the previous section has some rather drastic implications for how R & D subsidies should be organized. The idea that a subsidizing agency should be able to "confiscate" the entire profit of a subsidized project will probably sound quite alien to any political body deciding whether to institute such a policy. Our intention is merely to derive some criteria from the theoretical ideal that can be used to judge the effectiveness of subsidy policies that are practicable and politically possible.

The first principle that is clearly embodied in the incentive subsidy is that firms should retribute the subsidizing agency in proportion to the profit it earns from a subsidized project. A particularly important point is that the size of the retribution should not be limited to the size of the subsidy. If a firm earns large profits from a subsidized project it should be obliged to repay a share of the profit that is larger than the subsidy it received. The reason that this point is so important is that if the repayment never exceeds the amount of subsidy a firm will always expect to gain or break even when applying for a subsidy. There is no incentive for firms not to apply with projects it would have conducted anyway. In contrast, if there is chance that the share of profit is made to repay is larger than the subsidy the firm will not apply with many projects that are expected to turn in a profit. Thus the self-selection of firm applications becomes effective only when there is a chance that the repayment is larger than the subsidy.

The second principle embodied in the incentive subsidy is that there is an insurance component. The firm is not required to repay if the project does not earn a profit. This principle is important because it induces firms to conduct projects that have positive private expected values but that are too risky. This insurance is also advantageous from a fiscal point of view. It contributes to the policy goal of raising the level of R&D and yet it may be on average virtually costless to the public purse, provided that the repayments from those that succeed are sufficiently large to cover the losses from those that fail.

The third principle is that efforts that the firm makes to raise social value rather than private value should be rewarded. In many cases raising social and private values may require the same decisions. In those cases this third principle can be ignored. It is important however when social and private aims do not coincide. As an example, often the diffusion of research results lowers the private value of an invention but raises the social value. In those cases a reward for efforts to disclose the results of research may be appropriate. Another example is that rewards may be in place for designs that reduce environmental hazards.

The three principles are summarized below:

1. Retribution of subsidy
2. Insurance against failure
3. Incentive to take consideration of social aims.

These three principles can be used to reassess the list of subsidy instrument. This list is reproduced from chapter 2.

Table 7.2 A list of subsidy instruments

General subsidies
   1. Tax deduction for R&D expenses
   2. Tax deduction for a rise in R&D expenses
   3. Personnel grant toward costs of R&D personell

Selective non-self-financing subsidies
   4. Project grants
   5. Project loans at subsidized interest rates
   6. Conditional loans that are repaid only if R&D is succesful
   7. Loan guarantees
   8. Prizes

Selective self-financing subsidies
9. Fee-based loan guarantees
10. Royalty grants. Royalty to the state is based on sales of the invention toward which the grant was applied.
11. Stock option grants. In return for an R&D grant the state receives a stock option that can be exercised if the stock value rises significantly. For large firms the stock option refers to separate venture companies set up around the respective R&D project.
12. Convertible loan. The state gives a loan that can be converted into stock if the project turns out to be a commercial success.
13. Equity investments. The state invests directly or via private investment companies in venture firms.

It is quite clear that all the general subsidies fail the first and the third principle. There is no repayment required for general subsidies and there is no reward for better fulfillment of social aims. The second principle is fulfilled in a weak way. General subsidies reduce the cost of researching and this cost reduction need not be repaid regardless of whether the project fails or succeeds. The insurance effect is small however because the cost reductions that general subsidies imply are small. This is a consequence of the subsidy budget being spread thinly over a large base of projects.

The non-self-financing subsidies fulfill the first two principles somewhat better. Some of these subsidy instruments have repayment provisions although they do not allow for repayments that are larger than the subsidy was.

The self-financing subsidies with the exception of the fee-based loan guarantee fulfill the first two principles rather well. They correctly demand a retribution of the subsidy and they contain an insurance component. The third component can always be added to the scheme, providing some reward for fulfilling social aims.

To summarize, we have derived some theoretical principles for the design of efficient subsidy instruments. Using these to evaluate existing subsidy instruments yields the result that those instruments that we on theoretical grounds would expect to be most efficient by and large are those that in empirical investigation turn out to be most efficient.

7.4 Subsidy targets and rules

Efficient subsidization is not just a question of choosing the most efficient instruments but also of choosing those subsidy targets for which subsidies can be most effective. In this section we discuss a number of such choices one can make to raise the efficiency of subsidies.

The first choice, suggested by the empirical evidence – e.g. the Swedish study – concerns the size of firms. Apparently small firms are much more responsive to subsidies than large firms. Presumably this reflects the fact that small firms often face a financing constraint that limits the amount of R & D they can conduct.

A second, more difficult, choice is to what extent one should subsidize projects for which the social value is large in relation to the private value. For example, should one subsidize R & D leading to products that are profitable for the firm, or should one subsidize R & D which may be less profitable to the firm but have a large social value such as environmental technology. This choice often embodies the following trade-off: Concentrating on projects that lead to results profitable to the firm raises the risk that one subsidizes projects that the firm would have conducted anyhow. On the other hand concentrating on projects that are less profitable to firms raises the risk of subsidizing projects that for technical or economic reasons are poor choices. In effect the latter alternative places a greater burden on the subsidizing agency of defining worthwhile projects. If the subsidizing agency has poor information relative to the firm then the latter approach risks a lot of mistakes. One may conclude from this that the best target group lies somewhere in the middle: projects that are rather risky and fulfill some social aims but still have some profit potential.

A further question is how much management support the subsidizing agency should supply. This is relevant mainly for subsidies to small firms. Venture capital firms have found that it pays to invest in thorough management support in the companies they invest in and to ensure that the firm has access to specialist competence. This suggests that subsidizing agencies should be doing the same for firms they subsidize. The problem is that a subsidizing agency may not be very good at management consultancy. There are however a wide variety of private consultancy firms that can be hired for this purpose.

Coordination of research

It is not popular in this day and age to propose that the government should coordinate private research. Yet two lines of thought raise the suspicion that private firms do not coordinate their research sufficiently in a way that may be ameliorated by government policy in some cases.

The first line of thought is that government coordinators' of research have had some spectacular successes. For example it is often argued that government supported military research has been much more successful than subsidized private research; the implication is that the centralized coordination of military research reaps efficiency gains that a case by case subsidization of private projects misses. Satellite communications, the transistor, and the computer are all inventions with major civilian value and yet they were, although conceived in the civilian sector, only made usable by aggressive and well-coordinated military research at a time when they received virtually no private or civilian government support (Schnee, 1978; Teubal & Steinmueller, 1982). Another example is Japan's coordination of research through its MITI agency (Oshima, 1984).

The other line of thought is theoretical. Clearly, firms competing towards similar inventions will duplicate some research leading to a waste of resources that a coordinator would have avoided. Even firms that do not compete directly are likely to keep some information secret that could directly benefit research done in other firms.

In sum it may be that the government should coordinate research in some cases or aid coordination. These cases arise when the costs of bureaucracy are smaller than the losses due to lack of coordination.

In these cases the government may play a role as a coordinator and a guarantor for each firm getting a fair share of the returns in accordance with what it put in.

The question is how a coordinator can get firms to reveal their ideas truthfully and then to follow the coordinator's instructions. Fölster (1986) treats this problem, suggesting an incentive compatible system of rewards and taxes.

Another possibility is to induce firms to arrange for coordination themselves. For example one can make the payment of subsidies conditional on firms coordinating their research as is done in the European ESPRIT programme where a firm gets 50% if its research costs subsidized providing it cooperates with another firm. Or one can encourage joint research ventures such as the MCC in Austin, Texas. Formed by 26 companies the research institute works on basic computer innovations. Each company sponsors a few projects in the institute. If a project succeeds the sponsoring firm gets exclusive rights to exploit the invention for 3 years. After that the inventions are opened up to general licensing with all companies dividing the spoils.

8 Summary of policy conclusions

After a rather lengthy discourse on various aspects of innovation policy it is time to sum up and formulate constructive policy proposals.

A good portion of this book is concerned with empirical measures of the effectiveness of innovation subsidies. A fair summary of this discussion is that some of the most common subsidy instruments do not prove to be very efficient. They do not generate a lot of additional R & D in relation to the size of the subsidy. Cost-benefit studies like the one illustrated in chapter 1.3 suggest that subsidies may not be worthwhile unless efficiency can be raised.

If the most common subsidy instruments are inefficient why then are they used? The simple answer seems to be that those designing and executing subsidy policies did not know better. It is a fact that subsidizing agencies have not shown much interest in evaluating the efficiency of innovation subsidies in a serious way. Further, a number of case studies and our experiment on project selection (chapter 4.3) arouse the suspicion that subsidizing agencies often do not even have a good grasp of the basic principles that determine whether a subsidy is worthwhile from a social view or not.

We have attempted to explain subsidy agencies' lethargic attitude toward policy evaluation with reference to the political situation they are in. This can be described as follows. Serious policy evaluations can be performed only with the cooperation of subsidizing agencies. As long as such evaluations are not performed subsidizing agencies are in a comfortable position. Their "customers", the firms receiving subsidies, are not likely to complain unless, perhaps, the administrative burden of applying for subsidies becomes too large. The government on the other hand has nothing to complain about since it cannot point to any obvious mistake or lack of organization on the part of the subsidizing agency. It can not even point to foreign subsidy agencies since they operate roughly on the same principles.

After this brief summary of the main criticism we turn to the policy proposals. They begin with an appeal for a "politically" independent organisation in charge of policy evaluation. Then they turn to the specific methods that should be used for policy evaluation. Finally, we have suggestions for the organization of project selection and the choice of subsidy instruments.

Since we have argued that subsidizing agencies often have been uninterested in policy evaluation the logical solution is to let an independent organization take charge of evaluation. This follows the proven principle that conduct and control should not be concentrated in the same hands. Such an independent organization could be a state agency. Alternatively independent evaluation experts could be contracted. They should, in the event, be contracted by a part of the government that has no vested interests in innovation policy such as the ministry of finance.

Because the "evaluation agency" in some sense has interests that conflict with those of the subsidizing agency it is important not to recruit its employees entirely from the vicinity of the subsidizing agencies even though those persons may be most knowledgeable about the operation of current subsidy programmes. Instead the evaluation agencies should employ a fair proportion of social scientists who have a good grasp of the empirical methods necessary for policy evaluation as well as of the economic principles underlying the calculation of the social value of innovations. The evaluation agency should make a point of not relying too much on "soft" evaluation methods based on interviews with subsidy administrators.

Apart from pure evaluations the evaluations agency serves as advisers to subsidizing agencies. Thus it can provide the legal and financial expertise that the subsidizing agencies may need occasionally to apply specific subsidy forms. In particular those subsidy forms that entail some pay-back provision require a sound legal backing.

If the evaluation agency is chosen to be a public agency it could also perform a number of coordinating tasks that today are neglected in most countries. The most important of these is to provide a common database for all the different subsidizing agencies. In this database a subsidizing agency could check instantly whether an applicant has received subsidies from others, who the main competitors are to a specific technology, who the experts are in various technological fields that are capable of judging the merits of a project and so forth. The evaluating agency needs such a database anyway to prepare summary statistics and it would be a simple matter to give subsidizing agencies access.

A further coordinating role that the evaluation agency could serve is as a common bank. Today the prevailing system is that evaluating agencies receive budgeted funds that they then are eager to spend. There is little opportunity for reallocation during a budget year if it turns out that one subsidizing agency receives much more promising subsidy application than another agency. If the evaluation agency served as a common bank this problem could be ameliorated. Subsidizing agencies could then request funds at the rate at which they receive interesting applications. The evaluation agency could continuously signal how good applicant projects have to be to receive further funding. Thus one could achieve a rough reallocation between subsidizing agencies.

At this point it may appear that a major issue has been circumvented: Who decides how much each subsidizing agency can disburse? As described above the evaluation agency has a measuring and coordinating role but it is not a central decision maker. Rather the government budgets funds to each subsidizing agency just as is common practice today. The new elements are first that the government makes its decision under consideration of the evaluations performed by the evaluation agency and, second, that it provides a possibility for subsidizing agencies to release funds that the evaluation agency then can reallocate to other agencies.

The crucial aspect of this system is that subsidizing agencies must prefer to release funds rather than being criticized by the evaluation agency for granting subsidies that are not socially worthwhile. The government can achieve the proper incentives by threatening to close down subsidy agencies that consistently allocate subsidies poorly. In contrast an agency that often releases budgeted funds may receive a cut back in budgeted funds but may survive with its name in good repute.

The threat of closing down subsidizing agencies may have to be executed occasionally. In turn one can allow new ones to begin operating. As argued below it is even possible to funnel funds through private investment- and venture capital companies.

Policy evaluation methods

Policy experiments are the most effective method for evaluating policy effectiveness. Many current policy programmes can be reorganized as experiments with only slight alterations in the way they are executed. The main considerations in designing experiments were discussed in the previous chapter and a detailed example of a policy experiment is given in chapter 11.

Previous experience with policy experiments shows that they can easily fail because political pressures lead to biases in the conduct of the experiment. Therefore it is important that the responsibility for the experimental design rests not with the subsidizing agency but with the independent evaluation agency.

The experiments should be complemented with the traditional tools that also have been discussed at length above: Surveys and econometric analyses.

All of these empirical techniques should be applied within a cost-benefit framework, carefully spelling out what the social costs and benefits of a subsidy are. Even when it is not possible to quantify these costs and benefits with any precision the subsidizing agencies should have a clear understanding of the trade-offs involved.

The results of these studies will not always be clear cut. Occasionally misleading results may lead to mistaken decisions. One should keep in mind however that the purpose of policy evaluation is to improve decision making on average. Thus it may not be possible to eliminate mistakes entirely, only to make them less frequent than under the current system.

Subsidy instruments

In the previous chapter we have at some length discussed the principles that subsidy instruments should adhere to in order to be effective. In practice different situations require different subsidy arrangements. Our general policy proposal is to build these principles into the various subsidy instruments one uses.

Both from the theoretical discussion and the empirical results the most effective subsidy instruments are those that give the public purse a share in profits in return for the subsidy. Examples of such subsidies are stock option grants or equity investments. These instruments are equivalent to investments that private investors conduct when they buy shares in a company. Administratively the stock option grant is different because it avoids public ownership of shares. For most practical purposes however it is quite similar to ordinary stock investment.

To those who believe in the efficiency of free markets it may come as no surprise that this subsidy instrument turns out to be efficient. After all it is very similar to the instrument that private investors have found to work best.

There are some limitations to the use of stock option grants. For example, they are difficult to use when subsidizing a project in large firms that refuse to break the project out of the firm and form a separate company. However the empirical study suggests that all kinds of subsidy instruments work less well with large firms. Thus it may not be as worthwhile to subsidize R & D in large firms as in small.

Public equity investments or stock option grants can even be funnelled through private investment companies. This is desirable to the extent that private investment companies are subject to tougher selection mechanism. A poorly functioning private investment company eventually goes bancrupt or changes management or owner. As a result only the most successful continue.

We know from experience that this filtering process works less well in public agencies. Governments and parliaments react more slowly to incompetent management or changing conditions. Agencies cannot be acquired by others who believe they can do a more successful job.

A solution can be to let private venture capital companies invest public funds in R & D intensive start-up firms. The public funds are lost if the investments fail. If the investments succeed the public purse receives a return corresponding to the share of public funds invested.

With this construction the private investment firm has incentives to use public funds for projects that they consider too risky to use their own money for, but that they would like to promote up to a point were they seem less risky. At that stage the private investment firm will invest own funds and thus become eligible for a share of the profit. The cost to the investment firm is that it has to use its expertise and time to select interesting projects. In return it gets a first hand right to invest itself once the project have grown less risky and starts to look like good investments.

This system has several other advantages. One is that evaluation of policy effectiveness becomes relatively easy since one can compare the performance of different investment companies. Allowances can also be made for projects that have social values beyond their expected profits.

Part four: Selected studies

9 The efficiency of innovation subsidies

9.1 Introduction

State subsidies to R & D or innovative investments in firms are organized in many different ways. Examples from the plethora of extant subsidy instruments are tax incentives, grants to researchers, project grants, loans, conditional loans, and grants with royalty rights. Very little is currently known about the effectiveness of these subsidy forms.

In this chapter we compare the effectiveness of eight forms of subsidy for R & D projects. The comparison is based on a survey of Swedish R & D managers, including detailed information about 214 research projects or project proposals. In a first set of results we report managers' general judgements about the effectiveness of different subsidy forms. Second, R & D managers were asked to judge how each subsidy instrument would affect the firm's decision about the size of each project and whether to conduct it. This allows an estimate of how much additional R & D each policy might induce.

There are two main conclusions. First, general subsidies do not seem to induce much additional R & D for a given amount of subsidy. Second, among specific subsidies so called "stock option grants" seem to induce most R & D per subsidy krona. These are grants that give the state a right to recoup some of its funding by exercising a stock option if the firm's value rises rapidly. The main reason that the stock option grant performs well is not that the state can recoup some of its costs but rather that firms do not accept this subsidy for much of the research that they would have conducted even without subsidy.

In theoretical models the "efficiency" of subsidies is easily defined as the change in some social welfare function. For empirical purposes, however, efficiency has usually meant how much additional R & D is generated for a given cost to the public purse. Undoubtedly this definition ignores a number of efficiency aspects such as the extent to which the conduct of R & D is adversely affected by the subsidy application procedure and subsidy regulations. Nevertheless it probably captures the central element and it is tractable empirically. Thus in the following we take the term efficiency to mean the additional R & D generated for a given outlay.

A few previous empirical studies have endeavoured to estimate the efficiency of different subsidy instruments. These are reviewed at length in chapter 4. Three different empirical methods are used. One is the case study. The other are econometric estimates of the correlation between subsidization and R & D intensity across industries or firms. The third method consists of surveys. All of these studies concern one or other existing policy, and in no case, as far as we are aware, is an attempt made to compare the impact of different subsidy instruments on similar projects.

Case studies (e.g. Roessner, 1984) always leave the question open of how representative the studied cases are. The econometric studies (e.g. Lichtenberg, 1984) have to date not been able to convincingly discern the direction of causality in the correlations between the amount of subsidy and the amount of firm R & D spending. A common finding is that total R & D expenditure is larger in industries that receive subsidies, but the difference in R & D expenditure is smaller than the amount of subsidy. Such correlations can be explained either by the fact that subsidies stimulate R & D or by the fact that firms receive greater subsidies if they have promising research ideas and, therefore, greater incentives to invest themselves. As a result of this problem our judgement is that survey methods are as likely to produce useful answers as econometric studies are.

The survey studies fall into two groups. One approach has been to query respondents about their general judgements concerning a policy. The other is to focus on specific decisions and ask how they would have been changed in the presence of a policy (e.g. Gronhaug & Frederiksen, 1984; Mansfield, 1986). In this paper we do both. This provides a control of the extent to which respondents merely draw on their general judgements when they reconsider specific decisions. The specific decisions in turn permit a quantitative estimate which is necessary for a judgement of whether subsidies are socially worthwhile.

Our survey among Swedish firms includes projects that firms conduct as well as projects that firms for the time being have decided not to conduct. Roughly half of the firms were large firms with more than 100 employees (571 employees on average). The other half were venture firms with 24 employees on average.

The subsidy instruments we analyse are listed in Table 9.1. They can be divided into three categories: General subsidies, selective self-financing subsidies, and selective non-self-financing subsidies. Selective subsidies are those that are approved on a case by case basis. Self-financing subsidy systems include repayment provisions that make it theoretically possible that the subsidy program as a whole will be self-financing. Whether these systems actually are self-financing in practice depends of course on the exact provisions and on the projects that are subsidized.

Section 9.2 describes the survey design. Section 9.3 reports the research managers' general judgements of policy effectiveness. In section 9.4 the quantitative estimates of policy effectiveness are shown.

Table 9.1 Subsidy system

General subsidies
    1. Tax deduction for R & D expenses
    2. Grant toward costs of R & D personnel

Selective non-self-financing subsidies
    3. Project grants
    4. Project loans at low interest rates
    5. Conditional loans that are repaid only if R & D is successful

Selective self-financing subsidies
    6. Fee-based loan guarantees
    7. Royalty grants, royalty to the state is based on sales of the invention toward which the grant was applied.
    8. Stock option grants, in return for an R & D grant the state receives a stock option that can be exercised if the stock value rises significantly. For large firms the stock option refers to separate venture companies set up around the respective R & D project.

9.2 The survey design

The survey was carried out via personal- and telephone interviews. Such interviews rather than a questionnaire were deemed necessary because the questions were relatively complicated. Early trial runs indicated that respondents needed a fair amount of explanation in order to be willing or able to answer.

The questions were designed with guidance from the theoretical literature on subsidization and on technological change and the empirical literature on the efficiency of R & D subsidies. The interviews were held with high-level R & D managers, usually with responsibility for the R & D of a business unit.

The R & D managers were asked to report typical experiences or central tendencies within their line of business. They were thus treated as informed observers of the industry. In addition they were asked to select a number of representative R & D projects and were asked specific questions about these projects.[17] Some of these projects had been rejected and were not actively pursued. Respondents were asked to pick rejected and accepted projects in about the frequency

---

[17] A project is defined to be a fairly well-specified research proposal that can be accepted or rejected without significantly affecting the remainder of the firm's research activity.

with which they were proposed. For example an R & D manager rejecting about half of all well-defined project proposals would be asked to answer questions about an equal number of accepted and rejected projects.

Respondents were told that they need not divulge the technical nature of the projects so there was no reason for them to give misleading replies in order to protect secrecy.

Sample construction

The total sample consists of 61 respondents. Of these 33 are R & D managers of large business units with more than 100 employees. 28 are managers of small, newly started, firms usually organized around a single product or line of business. Each of the R & D managers of large business units gave details about 3–5 research projects. The managers of small firms gave details of two or three projects. In total the number of projects in the sample amounts to 214, of which 135 come from large firms and 79 from small firms.

Firms were chosen so as to make the sample representative of Swedish industry with one important caveat. Firms that do not conduct R & D were excluded. In total the sampled firms conduct about 6% of Swedish private R & D. No projects that currently receive subsidies were included.

Methodological issues

Because of the small number of firms in each industry we do not attempt to distinguish between industries. Some of the variance in the data may be explained by industry differences although none of the differences in evaluations of policy effectiveness are statistically significant between industries.

There is considerable variance in judgements of policy effectiveness between projects, even within each firm. This is reassuring because it means that respondents did not indiscriminately apply their general judgements to specific projects.

In the first part of our survey managers were asked about their general judgements of the effectiveness of the policies. The answers are reported on a seven-point Likert scale ranging from "not at all effective" to "very effective." There is no natural or objective anchor for such evaluative ratings. Individuals may perceive the same environment but simply use the scale differently. Some might systematically favor high scores; others might concentrate responses in the center of the scale. A number of techniques are available to control for differences among respondents in mean and variance. These techniques however impose the restriction of assuming a "true" uniform mean or variance. Rather than impose such restrictions we let the second part of our survey that depends on quantitative estimates rather than semantic scales act as a test of robustness.

Survey results are often biased by the ordering of questions. To avoid this problem we randomized the order in which questions were asked.

9.3 General judgements of policy effectiveness

R & D managers were asked how effective they believed different subsidy instruments to be in terms of stimulating additional private R & D at the lowest cost to the public purse. Respondents were asked to rate their judgement of effectiveness on a 7-point Likert scale ranging from 1 (not at all effective) to 7 (very effective).

Table 9.2 reports the results. The first two columns show the overall sample means for large and small firms respectively. The results are robust to the use of alternative summary statistics such as the median.

Table 9.2 General judgements of subsidy effectiveness

	Sample means		All firms
	Large	Small	
1. Tax incentive	2.1	3.2	2.5
	(0.11)	(0.13)	
2. Grant to R&D personnel	2.4	3.1	2.5
	(0.12)	(0.13)	
3. Project grants	2.8	3.3	3.0
	(0.10)	(0.12)	
4. Project loans	2.5	2.9	2.3
	(0.13)	(0.14)	
5. Conditional loans	3.0	3.5	3.3
	(0.11)	(0.11)	
6. Fee-based loan guarantees	1.5	2.2	1.8
	(0.14)	(0.13)	
7. Royalty grants	3.2	3.9	3.5
	(0.16)	(0.18)	
8. Stock option grants	3.6	4.2	3.9
	(0.11)	(0.12)	
All policies	2.6	3.2	2.8

Range 1 = not at all effective; 7 = very effective; Standard errors in parentheses.

There is a clear pattern in the results. Apart from fee-based loan guarantees the self-financing instruments are generally rated higher than non-self-financing instruments. In particular the general subsidies were rated low. Interestingly a number of managers commented that, if given a choice, they would prefer general subsidies even though they did not believe these to be an effective way of raising the level of R & D. Apparently managers had no difficulty in distinguishing between the firms' interests and the public interest.

Stock option grants were rated highest for both categories of firms.

In follow up questions we asked managers why they rated subsidy instruments in the way they did. We cannot report all responses here. Rather we summarise the comments that were shared by at least 20% of the respondents.

1. General subsidies were thought to attractive due to their administrative simplicity. They were thought to be rather ineffectual, however, because the thin spread of subsidies to all research means that the impact on any particular project is small.

2. Small firms were thought to be in greater need of capital. Thus subsidies to small firms were thought to have a greater effect. An additional consequence is that grants have the advantage over loans of not affecting small firms' already extended leverage.

3. The fee-based loan-guarantee scheme was viewed with suspicion. It was thought that unless it contained a large subsidy component it would be taken up largely by those already planning to default.

4. The stock option grant and royalty grant was thought by many to be attractive because "it resembles what private investors do". Since firms initially receive a grant their leverage is not affected, and the self-financing component is activated in proportion to the success of the project. Therefore these instruments were thought to reduce risk effectively while at the same time providing the state with a way of recouping costs.

9.4 Judgements of policy effects on specific projects

The research managers' general judgements shown above provide some insight. It is quite unclear however how robust they are. Further it is unclear whether, in the absence of a quantitative estimate, the stimulative effect of subsidies is large enough to justify their social cost.

In order to make quantitative estimates each R & D manager was asked to choose a number of representative R & D projects, including some that the firm had decided not to conduct at the moment. It was stressed that the ratio of conducted to non-conducted projects should approximate the proportions in which projects actually ocurred in the firm.

For each conducted project managers were then asked, for one subsidy instrument at a time, whether they would apply and to what extent the receipt of a subsidy would raise the firm's investment in the project. For each non-conducted project managers were asked whether they would conduct the project under each subsidy scheme and how much they would invest.

To be meaningful these questions require an exact definition of the size of the subsidy under each system. The conditions for each policy were specified in ways that roughly correspond to policies that actually exist. An additional consideration was that the total public expenditure implied by the subsidies should be as equal as possible. Since the definitions had to be fixed a priori it was of course not possible to align public expenditure exactly. Table 9.3 shows the exact subsidy specifications.

Table 9.3 Definition of subsidy systems

1. Tax incentive: 30% of R & D costs can be deducted from taxable firm income.
2. Grant to R & D personnel: 20% of the wages of R & D personnel are paid.
3. Project grants: 50% of project costs are paid.
4. Project loans: 70% of the project costs can be borrowed at a zero interest rate.
5. Conditional loans: 70% of the project cost can be borrowed at market interest rate and need not be repaid if the project fails. Failure means that the invention is not used or sold.
6. Fee-based loan guarantees: For a fee of 2% (large firms) or 5% (small firms) of the size of the loan 100% of the project cost can be borrowed at market interest rates. In case of bancruptcy the state picks up the loan.
7. Royalty grants: A grant of 50% of the project cost is given in return for royalty payments worth 5% of total revenues on the new product.
8. Stock option grants: A grant of 50% of the project cost is given in return for an option to purchase stocks within the next ten years at current prices and a volume of stocks corresponding to the amount of the grant at current stockprices. In large firms a separate venture company is formed around the project and the stock option refers to this venture company.

Our results about the effectiveness of subsidy instruments necessarily refer only to the exact specification of the instruments as shown above. This is unfortunate in the sense that a subsidy instrument that we find to be inferior to another actually may be superior with a different specification. This opens considerable scope for further research. One would expect however that if a subsidy instrument had dramatically different effects with a different specification then this should be reflected in the general judgements reported in the previous section. We take the fact that the general judgements coincide fairly closely with the quantitative effects as evidence that the subsidy instruments display similar efficiency even with different specifications.

Another consequence of using exact specifications is that the total public cost of each subsidy system cannot easily be held equal for all subsidy instruments. In particular for the general subsidies the total public expenditure is determined entirely by the managers' responses. For the other instruments, however, it is possible to fix the total budget and grant the subsidy to as many projects as the budget allows. Thus, given the managers' responses one can manipulate one policy parameter, the total budget, even though the size of the subsidy per subsidized project cannot be changed.

For some of the policy instruments additional questions had to be asked to determine the size of public outlays required. These questions and the exact procedures for calculating public outlays are reported in the appendix.

In table 9.4 we show with how many projects firms would have applied to each of the subsidy instruments. In general firms would have applied with most of the projects that they conduct anyway to the general and non-self-financing instruments. In some cases however firms reject subsidies. In follow-up questions managers indicate that in some instances they are worried about maintaining secrecy about projects when applying for a subsidy to a public agency. In other cases the subsidy instrument does not work. In particular the tax incentive is not taken up by all firms because it only represents a subsidy to firms that earn a profit.

The self-financing subsidies are taken up much less frequently for projects that firms would have conducted anyway.

Of the projects that firms do not currently conduct the firm would accept the subsidy for some fraction of projects and would then be willing to conduct the projects.

Table 9.4 Number of projects for which a subsidy is sought, in percent of conducted and non-conducted projects

	Conducted projects		Not conducted projects	
	Large	Small	Large	Small
1. Tax incentive	95	71	10	8
2. Grant to R&D personnel	100	100	13	9
3. Project grants	91	97	22	25
4. Project loans	87	96	19	21
5. Conditional loans	87	97	17	23
6. Fee-based loan guarantees	2	15	0	5
7. Royalty grants[a]	32	34	18	29
8. Stock option grants	14	29	19	23
Total	63	67	15	17

[a] Only projects resulting in products were applicable to royalty grants. These were 55% of projects in large firms and 68% of projects in small firms. Here the percentage of applicable projects is shown.

To provide a proper comparison of policies we must simulate the selection of projects that receive selective subsidies. We assume that the general subsidies are granted to all firms that apply. The selective subsidies are applied only to a subset of projects selected from all projects that firms say they would apply with. This selection process essentially expresses how accurately the subsidizing agency can distinguish projects that should be subsidized from those that should not. We examine three levels of information that subsidizing agency might have:

1. *Perfect information:* Of all projects that apply only those projects receive a subsidy that either would not have been conducted without the subsidy or where the investment in the project is increased by at least half the amount of the subsidy.

2. *Inperfect information:* Half of all projects are selected as with the perfect information criterion. The other half are selected as though the state had no information at all so that all that apply receive the subsidy.

3. *No information:* All projects that apply are subsidized.

Tables 9.5, 9.6, and 9.7 show the amount of new R & D generated per krona of public expenditure for the three levels of information.

The general subsidies have the same effect in all tables since they are not affected by the assumptions concerning project selection. The general subsidies show relatively poor ratios of R & D generated to public expenditure.

The selective non-self-financing instruments perform fairly well under perfect information but with poor information they perform poorly. Since they are given indiscriminately with poor information one would expect them to perform similarly to general subsidies. Table 7 confirms this suspicion.

The loan guarantee is fairly insensitive to information levels. The reason is that so few firms apply to this scheme, in particular with projects they would have conducted anyhow. As a result this instrument may not appear inefficient in comparison with general subsidies but it certainly is ineffectual. Little new R & D is generated even though the costs to the public purse are not high.

The royalty grant and stock-option grant are also relatively insensitive to information levels. Again the reason is that few firms apply with projects they would have conducted anyway. As whole these grant systems, and particularly the stock-option grant, appears to generate most R & D per public expenditure.

Table 9.5 Ratio of R & D generated by the subsidy to present value of the subsidy with perfect project information

	Large	Small
1. Tax incentive	0.19	0.08
	(0.06)	(0.07)
2. Grant to R&D personnel	0.16	0.07
	(0.06)	(0.07)
3. Project grants	0.82	0.96
	(0.07)	(0.08)
4. Project loans	0.80	0.91
	(0.08)	(0.08)
5. Conditional loans	0.82	0.98
	(0.07)	(0.09)
6. Fee-based loan guarantees	0.74	0.61
	(0.005)	(0.008)
7. Royalty grants	0.92	1.12
	(0.11)	(0.13)
8. Stock option grants	0.99	1.17
	(0.09)	(0.10)

The standard errors are shown in parentheses.

Table 9.6 Ratio of R & D generated by the subsidy to present value of the subsidy with imperfect project information

	Large	Small
1. Tax incentive	0.19	0.08
	(0.06)	(0.07)
2. Grant to R&D personnel	0.16	0.07
	(0.06)	(0.07)
3. Project grants	0.41	0.52
	(0.06)	(0.07)
4. Project loans	0.4	0.59
	(0.05)	(0.07)
5. Conditional loans	0.47	0.64
	(0.06)	(0.08)
6. Fee-based loan guarantees	0.48	0.47
	(0.01)	(0.02)
7. Royalty grants	0.56	0.74
	(0.10)	(0.11)
8. Stock option grants	0.72	0.92
	(0.09)	(0.10)

The standard errors are shown in parentheses.

Table 9.7 Ratio of R & D generated by the subsidy to present value of the subsidy with no project information

	Large	Small
1. Tax incentive	0.19	0.08
	(0.06)	(0.07)
2. Grant to R&D personnel	0.16	0.07
	(0.06)	(0.07)
3. Project grants	0.21	0.30
	(0.05)	(0.06)
4. Project loans	0.18	0.27
	(0.06)	(0.07)
5. Conditional loans	0.21	0.29
	(0.06)	(0.07)
6. Fee-based loan guarantees	0.36	0.32
	(0.005)	(0.01)
7. Royalty grants	0.51	0.70
	(0.08)	(0.09)
8. Stock option grants	0.68	0.90
	(0.08)	(0.10)

The standard errors are shown in parentheses.

As a test of the robustness of our results one can compare them with an estimate of the elasticity of R & D with respect to research costs. To do this we asked firms what effect a cost reduction of 10% would have on each project. The response to that questions indicates an elasticity of R & D with respect to research costs of 0.26. This is in line with findings in previous research (e.g. Mansfield, 1986). It also fits well with our survey results. One would expect a R & D cost reduction to have a slightly greater effect on R & D than an equivalent general subsidy since the subsidy may be judged to be more uncertain.

Conclusion

A survey of research managers' reactions to hypothetical subsidies is used to compare the effectiveness of different subsidy instruments. The robustness of the results is confirmed by a number of checks. First, managers do not just give their general judgement but also judge how specific projects would be affected by the subsidies. Second, manager's judgement of the effect of hypothetical cost reduction reveals an R & D elasticity that is in line with the findings of previous research.

The main results are the following. The subsidy instrument that seems to perform best is a so called stock option grant. In general self-financing instruments seem to perform better and to be less sensitive to conditions of poor information. The only exception is loan guarantees that were viewed with considerable suspicion.

Appendix to chapter 9

The total public expenditure for each subsidy instrument is calculated as shown below. The survey contained questions about the project costs, number of employees, and duration that were used for all of the subsidy instruments:

1. *Tax deduction.* If the firm was earning a profit the public expenditure was calculated using the corporate tax rate that the firm had actually paid in the previous year. If the firm did not earn a profit the public expenditure was assumed to be zero. This means that we ignored the possibility of carrying over losses to future years.

2. *Personnel grant.* Here the public expenditure is simply a function of actual or planned R & D personnel and the duration of the project.

3. *Project grant.* Public expense is calculated as 50% of the project costs.

4. *Project loan.* Here the present value of the interest subsidy is calculated assuming a constant rate of inflation.

5. *Conditional loan.* Managers were asked how likely they thought that the project would be successful. Successful was defined as meaning that the R & D costs would be recouped. Then managers were told that they should expect to repay the loan with the same likelihood. Public expenditure was calculated using the likelihood that managers reported.

6. *Fee-based loan guarantee.* Here independent estimates of the likelihood of bancruptcy were used. These were derived from a sample of similar firms.

7. *Royalty grant.* Managers were asked the rough order of expected sales for product innovations. The royalty grant was applied only to product innovations. These estimates and project duration were used to calculate public expenditure.

8. *Stock-option grant.* To calculate public expenditure we make an extremely rough, but conservative, estimate of the value of the stock-option. In fact, with our assumptions the value of the stock option does not reduce public expenditure much. We assume that firms earn a total real profit of 2% (of R & D costs) a year on each conducted project. Then, assuming a constant p/e ratio we calculate how this would affect stock prices. For firms without listed stock prices we impute these using book values.

10 Performance measurement

10.1 Introduction

This chapter concerns ways of quantifying the social costs and benefits of research projects. It begins with a general argument about when such quantification is useful. Then section 10.1 discusses the cost-benefit principles that should be used in quantification. Section 10.2 contains a commented example of such a calculation. Section 10.3 describes a computerized cost-benefit model, called EPRO, that was designed to evaluate research projects in the energy sector.

Estimating the social value of public or private projects is thought by many to involve an immense calculation. Even so, and no matter how laboriously the information has been garnered, the estimates remain of dubious credibility. These calculations, also known as cost-benefit studies, often perpetrate controversial findings. Critics complain that biased assumptions, slipped in to guarantee convenient results, are hard to detect. Sometimes the quarrels have their roots in the sheer number of pages of such studies. Too often volume breeds inscrutability for all but the authors themselves.

Not surprisingly many have drawn the conclusion that it is impossible, or in any case not worth the effort, to estimate the social value of many projects.

This conclusion is unfortunate because it does not give cost-benefit methods a fair chance and because it has militated against efficient public decision making. Public agencies and enterprises, relieved of having to unite around a measure of their own performance, often remain vague about their performance goals as well. As a result public employees frequently pursue conflicting goals, sometimes at variance with the true business of the public sector: raising public welfare.

In this paper it is shown that rough estimates of social value can be produced fairly easily. Making these estimates explicit can sometimes help to make better judgements. They always make decisions much more transparent. In addition calculations using explicit estimates are an excellent pedagogical device.

A common argument is that even when some intuition regarding the range of a project's social value can be mustered, the uncertainty is so substantial that one cannot ascertain any number representing an expected value.[18] The analogous claim relating to the stock market would be that no stock price can be determined merely because a company's investments face an uncertain future.

Just as stock market investors can place a value on an uncertain asset, so ought a public decision maker be able to value the social return of uncertain projects. This need not require any more work or information than what is used anyway to make decisions.[19]

So not only is it in principle possible to place values on extremely uncertain outcomes, but in fact, it is argued below, research project administrators already do so implicitly. Turning the implicit estimate explicit can confer a significant advantage.

Anyone deciding whether to invest in a public project (or to subsidize a private project) can hardly avoid judging the project's merit. If such a project is turned down one has implicitly estimated its social value to be negative. If one considers a project to be a borderline case its social value is implicitly taken to lie in the vicinity of zero. To implicitly define the value of any project then one need only ask oneself at what level of costs one would consider the project a borderline case. To make the implicit estimate explicit simply subtract actual project cost from the hypothetical level of costs that makes the project a borderline case. The difference is the estimate of social value based on the information currently available to the decision maker.

This estimate may indeed be very uncertain. The important thing is that it is at least as well founded as the decision makers decision whether to support the project or not. Merely making the social value estimate explicit can, at virtually no extra expense, improve the quality of decisions and, perhaps more importantly, provide a yardstick for how efficiently projects are conducted.

The quality of decisions rises for three reasons: First, the explicit statement of social values allows experience to be communicated. For example, a less experienced decision maker can rapidly gain insights by observing how highly projects are valued by his more experienced peers. To illustrate this, suppose an apprentice observes how an

---

[18] The common claim that social value cannot be estimated at all is simply false. Some argue not only that the estimation of social values leads to suspect results, but also that it is impossible for logical reasons. This claim leads to an absurd conclusion. Suppose one is faced with a specific research project. Anyone claiming that nothing is known about the project's social value would have to conclude that the project has an equally large chance of contributing to social value as of detracting from it. The absurdity in claiming that nothing is known about social values becomes apparent when the project costs are either multiplied or divided by a factor of one million. In both cases the totally ignorant decision maker would still have to conclude that the project has a fifty-fifty chance of being worthwhile. This shows that even when the benefits of a research project are extremely hazy one usually does have some vague idea of the range of possible outcomes.

[19] Those upholding that stockprices are easier to determine than social values should be reminded that stock values are subject to speculative waves that are extremely difficult to forecast, probably much more difficult than any of the fundamental factors that determine social values. The fact that there are many agents simultaneously determining the stock price, instead of just a single decision maker, is of no relevance here since a speculator has to put a value on his own estimate of expected stock prices before he buys or sells. The meddling of other agents may simply complicate this task by generating unforeseeable speculative movements.

experienced administrator agrees to subsidize a series of projects. In the process the apprentice hardly learns to gauge the relative merits of the projects. Thus he remains ill equipped to decide on his own. Had the experienced administrator provided estimates of the value of the projects the apprentice could easily have detected the pattern and could have learned which projects to reject even when only observing accepted projects.

In addition, a kind of experience bank can be built that congregates administrators' judgements. This means that one decision maker can rapidly inform himself of how others have judged similar cases. In that way each decision maker receives feed back on how congruent his judgement is with others' evaluation. One might say that the explicit statements of estimated social value help to make tacit knowledge transferable.

Second, the quality of decisions may also be improved, because in calculating the estimates of social value a decision maker is made quite aware of what benefits a project must have in order to justify its costs. This may hone a decision maker's accuracy just as a marksman improves his aim by estimating distance, wind, and target movement. This point is more important than it might appear. Countless experiments show that experts in most professions, when confronted with information on which to base a decision, arrive at extremely poor configural judgements. A configural problem implies that a decision maker's interpretation of any single piece of information depends on how he evaluates many other inputs. This is typical of the skill required to judge the merits of research projects.

Ironically, as the experts in these experiments receive more information, their judgement often becomes poorer, but their confidence increases sharply. This well documented phenomenon demonstrates how crucial it is not merely to rely on intuition when making decisions, but to ascertain likely values for each input separately and to use available algorithms to determine the best configurations.[20,21]

The third way in which explicit estimates raise the quality of decisions is that the estimates can occasionally be tested more thor-

---

[20] For example in one study Slovic (1969) confronted a group of stockbrokers with eight important financial inputs (trend of earnings per share, profit margins, outlook for near-term profits, etc.) that they considered most significant in analyzing companies. The optimum solution could only be found in a configural manner. As it turned out, configural reasoning, on average, accounted for only about 4% of the decisions made. Moreover, the emphasis the brokers initially said they put on various inputs varied significantly from what they actually used in the experiment. Similar results can be found for example in Anderson (1972), Dawes and Corrigan (1974), and Goldberg (1971).

[21] A series of studies documents how managments' forecasts of earnings routinely miss the forecast by a wide margin. Examples are Copeland and Marioni (1972) and Basi et al. (1976).

oughly via in-depth studies of social values. These can then serve as a feed-back to administrators on the accuracy of their judgement.

Explicit estimates do more than just improve decision making: They can even improve the actual conduct of projects. As the social value of a project rises with increasing efficiency, any explicit estimate of social value contains an estimate of how well the project is being handled. This is important for three reasons. First, it provides the project manager with feedback about his performance. This avoids a recurrent problem, namely that contractors make wrong decisions merely out of uncertainty about what the contractee actually wants. Probably such a feedback has a motivational effect as well.

Second, a reward can be tied to an explicit estimate of social value. Consequently those conducting the project are likely to try harder.

Third, an explicit estimate of social value permits more complicated reward systems that can align the incentives of a subsidized firm with the aims of the subsidizing public agency. An example of such a reward system are the "incentive subsidy" and the some of the self-financing subsidy systems suggested in chapter 7.

Asking public employees to estimate social values differs very little from asking private employees to prepare profit estimates. Private firms estimate profit figures that are baked into their annual budgets. These are often extremely uncertain, but merely stating them explicitly allows the different experts, say, the banks, managers, and consultants, to compare each others' forecasts, thus allowing them to learn from each other.[22]

The point is that businesses obviously believe that even very inaccurate measures to be better than none at all. Even mistaken estimates help to communicate tacit knowledge, and an inaccurate performance feed-back is better than none at all.

10.2 A simple scheme for estimating social value

To put teeth into the argument above the remainder of this chapter discusses how to prepare simple estimates of the social value of research projects. This section contains a conceptual discussion while section 10.3 gives a complete example.

Books have been written about the estimation of social values of

---

[22] Even profit calculations for the previous year are quite uncertain in private firms. This is true at the firm level, where calculated profit may vary substantially depending on the accounting method; but it is especially true at the product level where extremely crude measures are often used. For example the usual way of calculating product cost is to allocate overheads as a percentage of direct labour costs (e.g. Johnson and Kaplan, 1987). With labour costs now averaging 10 to 15% of total costs and falling, this procedure is quite misleading. The use of direct labour looks prohibitively costly when it is burdened by an 800% mark-up for overheads. This misleads because overheads are not necessarily driven by labour costs at all.

public investments. Some of the associated problems such as choosing a discount rate or evaluating external effects have been discussed at length. In contrast very little has been written about the evaluation of R & D projects.

Research and development projects or results must be evaluated with due attention to the fact that their end product, knowledge, has some peculiar traits. For example inventing even the grandest contraption is worthless from a social perspective if someone else already has it or is about to invent a close substitute. Another example is that an invention's social value may be greatly diminished if its diffusion is delayed, say, by the monopolistic ambitions of the patent winner.

The more diffuse the benefits of research are, the less sense there is in trying to estimate all component parts in detail. In many cases however it is possible to ascertain a better view of the factors that determine social value. When this can be done it is much preferable to simple intuition because, as shown above, man's poor configural judgement plays tricks with our capacity for accurate assessment. Using exactly the same information greatly improved decision making is often obtained by following an algorithm that identifies optimal configurations of components.

Consider the basic characteristics of such an algorithm, providing a small set of rules that lead to an estimate of social value even when information is poor. The idea is simply to add up estimates for the value generated within the firm, within other firms, and subsidiary effects such as spin-offs and education.

The first step in determining an R & D project's social value is to examine its value to the researching firm. Here one must be careful only to add in extra profits expected by a firm due to the research results. For example if the research project leads to an improved product one should count only the expected profits from selling the new product minus the foregone profits of selling the old product. Likewise if there is another new technology that is a close substitute one should discount the expected profit by the probability that the substitute works, rendering the first invention useless.

When the outcome of a project is uncertain the potential profit should be discounted by the probability of achieving it. Two things are noteworthy here. First, society as a whole is virtually risk neutral, so no special allowance should be made for greater risk. Second, the social discount rate is generally considered to be somewhat lower than the private one due to the fact that society does not need to add a risk premium.

Next, add the project's external effects due to the spread of the invention. These external effects accrue to consumers, to other firms within the industry, and possibly to other industries. Finally add any other external effects. These may include environmental effects, educational effects, spin-off effects, and others. For example if a number of engineers learn some new skills that are useful in other areas these should be counted as social investments.

Estimating social values will never be a practical proposition until a fairly simple standard procedure exists for doing so. Such a procedure must undoubtedly center on standardized rules of thumb for approximating the different effects.

Ideally such a procedure should exist in a computerized version, attached to a database containing experience values for common external effects. Then one would merely respond to questions asked by the program which then automatically chooses the correct experience values, calculates the estimate of social value, and possibly, stores the answers to amend the database.

As an example, the calculation of social value could follow the algorithm below. The algorithm consists of determining the following 10 terms. Example values are shown here for illustration. Further below it is explained how they are calculated.

1. Social values of a new product (PROD) 140
2. Social value of a cost saving (COSTS) 200
3. Multiplier of diffusion value to other firms (DIFF) 5
4. Probability of success (PSUC) 0.50
5. Probability of a substitute (PSUB) 0.30
6. Time profile: Introduction after years (START) 3
   Increase percent Annually (INC) 10
   Peak year (PEAK) 10
   Decrease percent Annually (DEC) 30
   End year (END) 15
7. Project costs (CO) 90
8. Educational effects (EDUC) 20
9. Spin-off effects (SPIN) 20
10. Environmental effects (ENV) .... − 10

The total social value is then calculated as the total value of the invention within all firms multiplied by the chance of success and of no substitute arising. This value is then discounted over the invention's life cycle before adding or subtracting costs and external effects.

(1) TOTAL PROJECT VALUE =
((PROD + COSTS) * DIFF * PSUC * (1-PSUB)) *

$$\sum_{i=START}^{i=PEAK} (1+INC)^{i-START}/_{(1+r)i} + \sum_{i-PEAK+1}^{i=END} (1+INC)^{PEAK} * (1-DEC)^{1-PE}$$

In the example this amounts to (assuming r = 0.03).

S = (140 + 200)*5*0.5*(1 – 0.3)* 11 – 90 + 20 + 20 – 10 = 4535

In the following it is explained how each of the categories ought to be calculated. For each category the theoretical principle is described followed by a rule of thumb that can be used for approximation.

1. The social value of a new product

Often the social value of a new or improved product can be captured better by trying to estimate the ensuing cost saving to the product user. Check the next section for the procedure to follow in that case.

When the user's cost saving is difficult to determine, for example in the case of a video machine, then one should try to grasp the value of the innovation to the consumer and to the producer. This can be illustrated in the following diagrams (partially adapted from Andersson, 1986).

Figure 10.1. Social value of a new product.

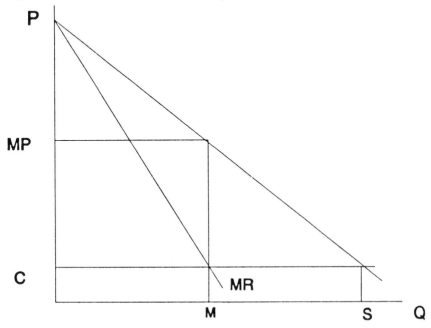

Figure 10.1 pertains to the case of a totally new product. It shows a demand curve for the new product. Much now depends on the pricing

Figure 10.2. Social value of a product improvement.

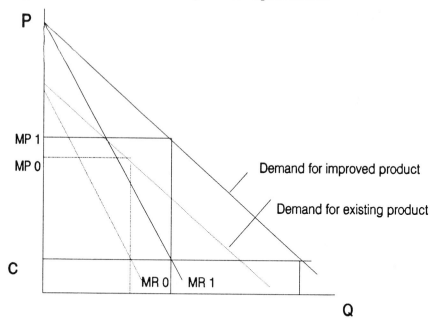

policy that the firm adopts. If there is no competition at all then it may adopt a monopoly price MP. The social value of the product is then the area under the demand curve up to M minus the production cost C * M. However if there is some danger of competition then the firm may price more leniently and the social value rises because the consumer surplus becomes larger by more than the private profit falls. From a social point of view it is preferable to set the price at the level of marginal cost C. This however leaves the firm empty-handed and lacking incentive to research.

Figure 10.2 demonstrates the principle behind valuing a product improvement. This figure shows the demand curves for the current product and for its improved form. The increase in social value corresponds to shadowed area between the demand curves up to the production level that the firm achieves.

Usually it will be unclear what the demand curve looks like, and perhaps even what the firm's pricing policy is. The following approximation will then serve the current purpose.

The demand curve one estimates should refer to the entire demand during the course of the first year of production. If it is easier to estimate the total demand over an uncertain range of years that may be preferable. Then one has to adjust the formula for the total social value to eliminate calculation of the production cycle for the innovation.

One can make a rough estimate of the slope of the demand curve, say very elastic (% change in sales/ %change in price = e = 3), unit

Figure 10.3. Social value of a cost reduction.

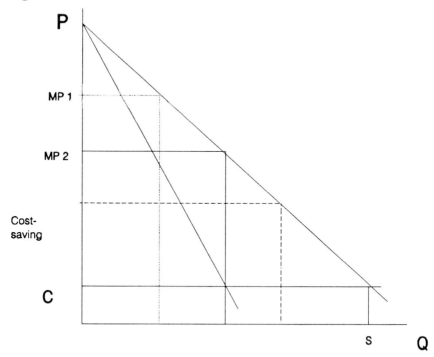

elastic (e = 1), very inelastic (e = 0.4). In addition one must estimate the unit production cost (C) and the likely sales price (PRICE) and the quantity sold at this price (Q). Then the social value can be calculated from the following formula:

(2) Social value of new product =
   $S(a_2 - C)/2 - (PRICE - C)(S - Q)/2$

The first term is the social value if the product were priced at production costs, while the second term subtracts a loss due to monopoly pricing. The demand curve is taken to be linear, $P = a_1 Q + a_2$, and the variables solve to

(3) $S = (a_2 - C) 1/a_1$
   $a_1 = e\, Q/P$
   $a_2 = PRICE - a_1 Q$

In the case of the product improvement one calculates the value of the new product as above and then one subtracts the value of the old product, also calculated by the same procedure.

2. The social value of a cost saving

The social value of a cost saving can be illustrated in the following diagram (adapted from Fölster, 1986). In Figure 10.3 again the demand curve for a product is shown. The cost reduction due to the innovation is now illustrated by a lower cost line. The shaded area represents the social value of this innovation, at least up to the production level chosen by the firm.

To approximate the value of this cost saving a procedure can be used that parallels the one used for valuing the product innovation.

The social value of a small cost saving is easily approximated. One can simply take the cost saving for the total current production. No account needs to be taken of any consumer surplus. However, when the cost saving is large the firm may lower prices. This raises demand. As a result the social value may be much larger.

If nothing is known about the shape of the demand curve then a reasonable approximation may be the following: Assume that the percentage of demand increase is inversely related to the percentage of price decline. Then calculate by how much the firm's sales would increase if it lowered price by the total cost reduction. Then multiply half of this demand increase with the cost reduction. In sum the social value is then calculated as (using the same notation as above):

(4) Social value = $Q*(C_{new} - C_{old})$
$(1 + 0.5*C_{new}/(C_{new} - C_{old}))$

where both total costs and the cost reduction refer to unit costs. If the likely cost reduction is large then it may be worthwhile to estimate the elasticity of the demand curve and take it into account as shown above in the estimation of the value of a new product.

3. Multiplier of value in other firms

In certain instances this multiplier may be easily determined. For example, if the invention benefits a clearly defined product in an equally distinct industry. Then one simply lets the multiplier equal the size of the industry relative to the size of the firm. This becomes more difficult if the invention has a variety of uses in different industries, or if it is unclear where the uses may lie.

The multiplier must also take account of the diffusion lag that arises before other firms use the invention. Since there exist quite a number of empirical diffusion studies one should be able to prepare experience values for the rate of diffusion.

4. Probability of success

It is a simplification to state a simple probability instead of a whole distribution. With that in mind the probability of success can be seen as simple weight reflecting the riskiness of a project. Specifically it will refer to the chance that the invention is marketed at all, or is used by the inventing firm. Given that the invention is used there is still a risk of how well it works (or sells). This risk should be baked into the estimate of the demand however.

5. *Probability of a substitute*

Here one has to take account of potential substitutes as well, in effect those that may be developed up until the time when the project is completed. If a perfect substitute is being developed then the project is not worth much. For example a project leading to a cure for cancer may seem precious now, but it would be worth much less in a situation where a cure by other means is already in the pipeline.

6. *Time profile*

This term only applies if the demand for the innovation stretches over a longer period. The measures have to include, first, an estimate of the time it takes to start producing and distributing, second an estimate of the development of demand, and third, an estimate of the speed at which substitutes or better inventions emerge, leading eventually to a decline in the use of the product.

7. *Project costs*

The project costs should be calculated by the net present value method, discounting properly for costs arising further in the future. Allowance should be made for possible "unexpected" costs that can arise even though they are not planned.

8. *Educational effects*

As a larger and larger fraction of firms' investments is made in knowledge rather than machines, it is important to recognize that even failed projects purvey knowledge that may come in handy elsewhere. To value this probably some simple rule of thumb is necessary equating the value of projects with the cost of equivalent courses. Thus a project lasting 3 years in which half the time was spent experimenting and planning may be counted as one-and-a-half years of education provided the knowledge gained has some relevance and can be used by the firm.

9. *Spin-off effects*

By definition the spin-off effects are unobservable at the time of evaluation. Thus any estimate in this area reflects a measure of how widely applicable the type of technology is. For example, an electronics component probably has a lot of spin-off effects, while a better mouse trap in all likelihood has fewer spin off effects.

10. Environmental effects

Here it is important to note that one should only count effects of the research project, not potential dangers that arise due to the fact that the use of the invention is not properly regulated. For example, if a project can leak a harmful new virus then this expected effect should be included here. However, if a virus is developed that is harmful if used improperly then the negative effects should not be ascribed to the research project, but to the use of the completed product.

10.3 The computerized cost-benefit program EPRO

EPRO is an application of the principles outlined above to the field of energy R & D subsidies. The program was originally commissioned by the Swedish Energy Board for application to a subsidy program that aimed to support the development and diffusion of technologies that save electricity.

EPRO can be divided into two steps. First, a conventional calculation of the private rate of return is conducted. This contains costs of investment, upkeep, expected electricity saving, and other cost components. In a second step the program estimates the expected value to society of subsidizing the project. The benefit to society is seen as the gain arising from a more rapid spread of the technology. Thus estimates how much faster diffusion may occur as a result of the subsidy and what the faster diffusion is worth. The value of faster diffusion is also called the "demonstration value" since the subsidy is often used to build a pilot plant that demonstrates the value of the technology.

An investment in untried technology runs a considerable risk of not generating any return. This risk can be large enough to scare off any investors. From a social perspective it may still be worth trying the technology however. If successful the technology would be imitated by many and thus generate a large social value. The first investor however cannot appropriate imitators' profits to himself. As a result the potential first investor undervalues the technology in comparison with its social value.

Figure 10.4 illustrates this problem. Figure 10.4a shows the uncertainty facing the first investor. Every possible outcome can occur with a certain probability.

Given a certain outcome for the first investor, imitators draw lessons and face less uncertainty. Figure 10.4.a also shows the second

Figure 10.4a. The firm's view of project return.

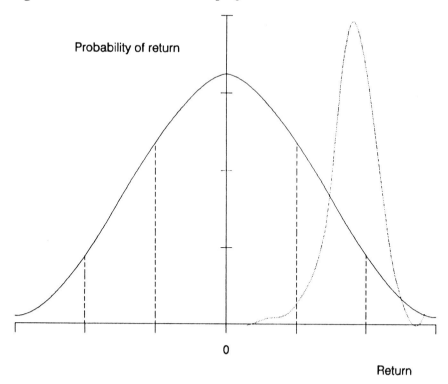

investor's uncertainty given that the first investor earned a large positive outcome.

For society a high positive outcome for the first investor implies a much larger social value than what the first investor earns. The corresponding probability function for society is therefore something like that shown in Figure 10.4b. Consequently society's expected value of an investment can be much higher than for the first investor.

This means that a subsidy should be granted in two cases. First, if the project has a negative expected private profitability but there is a chance that it turns out better than expected and in that case the social value is large. Second, the project has a positive expected private profitability but the uncertainty is so large that investors do not dare to invest.

EPRO handles uncertainty by demanding three values for each input variable. These are called the EXPECTED, HIGH, and LOW values. Formally the HIGH value is defined such that the risk of the variable lying above it is 5%. The LOW value is such that the variable has a 5% chance of lying below it. These values set a rough frame for the possible outcomes.

We will now demonstrate the use of EPRO with a very simple example. Suppose company A applies for a subsidy to test a new sort

Figure 10.4b. The social view of project return.

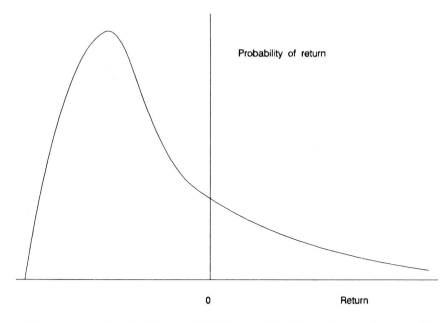

of insulation for its factory buildings. The best alternative to this project is to leave the factory buildings as they are. Costs are estimated to be the following:

The entire investment occurs during the first year. The new insulation is expected to cost 1.3 million. The high and low values are 1.8 and 1.2.

There are no costs of upkeep. Without insulation heating requires 400m^3 of heating oil per year. With insulation this is expected to fall to 350m^3, with the high and low values being 325m^3 and 375m^3. Oil costs 2000.- per cubic meter. The new insulation has a useful life of 25 years.

In addition the insulation saves some electricity. Without insulation 60000 kwh are used per year. With insulation this is expected to fall to between 40000 and 45000 kwh. The electricity saving occurs primarily during winter where electricity prices are higher.

With these specifications entered EPRO looks as follows.

Table 10.1 EPRO

PROPOSED PROJECT			YEAR 0	YEAR 1	YEAR 2
INVESTMENT					
	EXPECTED		1300		
	HIGH		1800		
	LOW		1200		
LIFE	MIN: 25	MAX 25			

UPKEEP				
	EXPECTED			
	HIGH			
	LOW			
FUEL				
	EXPECTED	0	700	700
	HIGH	0	750	750
	LOW	0	650	650
		YEAR 0	YEAR 1	YEAR 2
ELECTRICITY (kwh)				
	EXPECTED	0	42500	42500
	HIGH	0	45000	45000
	LOW	0	40000	40000
SUM				
	EXPECTED	1300	714	714
	HIGH	1800	773	773
	LOW	1200	662	662

BEST ALTERNATIVE		YEAR 0	YEAR 1	YEAR 2
INVESTMENT				
	EXPECTED			
	HIGH			
	LOW			
LIFE	MIN: 25 MAX 25			
UPKEEP				
	EXPECTED			
	HIGH			
	LOW			
FUEL				
	EXPECTED	0	800	800
	HIGH	0	800	800
	LOW	0	800	800
		YEAR 0	YEAR 1	YEAR 2
ELECTRICITY (kwh)				
	EXPECTED	0	60000	60000
	HIGH	0	60000	60000
	LOW	0	60000	60000
SUM				
	EXPECTED	0	820	820
	HIGH	0	830	830
	LOW	0	819	819

PRIVATE REQUIRED RATE OF RETURN:		20%	
SOCIAL REQUIRED RATE OF RETURN		6%	
PRICE OF ELECTIRICITY (SEK/KWH)			
EXPECTED	0.28	0.33	0.33
HIGH	0.42	0.50	0.50
LOW	0.26	0.31	0.31
DIFFERENTIATED EL PRICE			
STARTS YEAR	2		
PRICE WEIGHTS	1	2	4
% DISTRIBUTION	20	30	50
MARKET SIZE	100		
INITIAL USERS WITH SUBS	1		
INITIAL USERS WITHOUT SUBS	0		
TYPE OF PRODUCT			
CONSUMER			
COMMERCIAL	YES		
PLANT			

PRESENT VALUE WITHOUT SUBSIDY WITH SOCIAL DISCOUNT RATE

   EXPECTED   54
   HIGH   1047
   LOW   −1323

PRESENT VALUE WITH PRIVATE REQUIRED DISCOUNT RATE

   EXPECTED   −776
   HIGH   −225
   LOW   −1372

PAY-OFF TIME
   EXPECTED   12.3
   HIGH   7.6
   LOW   100

SOCIAL VALUE OF SUBSIDY   3157

RECOMMENDED SUBSIDY   776

The remainder of this chapter explains the technical side of EPRO.

Project expected value

The average of all cost entries are summed for each year. Based on this the difference between the project costs and the alternative's costs are calculated for each year. The difference is denoted X_t.

For each cost entry a variance is calculated with formula 4 below. The variances are then added. The expected value of the variance in any given year of X is v^2.

(5) $(HIGH - EXPEC)^2 + (EXPEC - LOW)^2$

According to the "central limit" theorem X can in some cases be considered normally distributed in spite of a certain variation in the probability distributions of the cost components. For a normal distribution 90% of outcomes lie within +/- 1.65 v of the expected value. Thus the expected, high, and low values are calculated as in formula (5). In most cases this approximation works well. A certain error can arise however, most pronounced when the calculation contains few cost components and the true probability distributions are extremely skewed.

(6)
$$\begin{aligned} \text{EXPEC } X &= X_t/(1 + 0.06)^t \\ \text{HIGH } X &= (X_t + 1.65 v)/(1 + 0.06)^t \\ \text{LOW } X &= (X_t - 1.65 v)/(1 + 0.06)^t \end{aligned}$$

EPRO makes adjustments when the project and the alternative have different lengths of life. In addition the program automatically performs a stepwise reduction of X between the highest and lowest length of life when this is uncertain.

Diffusion

The diffusion of technology is assumed to occur according to the following formula (6) of Lawrence and Lawton.

(7) $$N(t) = \frac{m + No}{1 + m/No \ e^{-Pt}} - No$$

N: Diffusion in year t
m: Number of users after diffusion
No: Number of users in the first year
P: Rate of diffusion.

A subsidy increases No which leads to a faster diffusion. Each year then the number of users that are due to the subsidy is calculated as N(t, with subsidy) – N(t, without subsidy).

The rate of diffusion is derived from empirical studies. It depends on the type of technology and how profitable the technology is.

The social value of diffusion

For each possible outcome X the value of the diffusion is calculated that occurs as the result of the subsidy during that year. The expected value of this over all possible outcomes of X is derived by numerical integration according to formula (8). Finally the present value of the stream of annual social values is calculated.

$$(8) \quad V_t = \sum_{0}^{\max X} X_t(N(t, \text{with subs}) - N(t, \text{without})) f(X_t) \, dX_t$$

V_t: Expected social value of subsidy in year t
$f(X)$: Normal density.

11 An experimental test of cost-benefit methods

11.1 Introduction

In most countries the public sector is responsible for a large number of investment decisions. Often these investments are merely financed or subsidized by the public sector but carried out by private firms or other organizations. In either case virulent controversy surrounds the choice of method used to evaluate the investment's merit. A few public agencies swear by sophisticated cost-benefit studies;[23] many in contrast detest anything but their own intuition, trimmed by experience; the majority however cater to both sides, commissioning an occasional cost-benefit study which they then feel free to ignore in decision making.

The experiments reported in this paper indicate first that traditional cost-benefit methods are in some cases worthless or even harmful to effective decision-making. Thus some of the scepticism among practitioners about cost-benefit methods seems to be justified. Second, it appears that a somewhat more sophisticated cost-benefit method that properly takes account of uncertainty can lead to more efficient decision-making than the intuition cum experience that practitioners generally rely on. These results emerge from a laboratory experiment and an experiment among actual practitioners of energy technology subsidy programs.

Considerable theoretical advances in the methodology of cost-benefit analyses have been made over the past three decades. Many of these have been packaged in the form of guidelines to practitioners.[24] In spite of these analytic and pedagogic efforts two disturbing facts are now obvious: Practitioners rarely use cost-benefit analyses; and there has been very little interest in testing whether cost-benefit methods actually improve decision-making.

Leff (1985, 1988) provides an illuminating description of a failed drive to introduce cost-benefit methods at the World Bank. Leff argues that practitioners refused to adopt cost-benefit because it did not meet their needs. In particular, the cost-benefit method ignored complicated inter-sectoral and inter-temporal effects. This made it difficult for practioners to determine the validity of the calculations.

One way of interpreting this experience, and similar experiences elsewhere, is that practitioners reject a method that does not provide

---

[23] We make no distinction here between the terms cost-benefit and social cost-benefit.
[24] Among the most widely used are Little and Mirrlees (1968) and UNIDO (1972).

a very good idea of how uncertain its conclusion is. The standard recipe for dealing with uncertainty in cost-benefit analyses is sensitivity analysis. Sensitivity analysis usually gives a good idea of the range of possible outcomes but a only a poor idea of the likelihood of different outcomes. Practitioners may therefore be perfectly rational in preferring intuitive decision-making, less exact to be sure, but with a better understood reliability.

Consider a conventional cost-benefit analysis of an investment project. Each variable that affects the return of the project is more or less uncertain. Furthermore probability distributions of the different variables may be correlated. The correct procedure would then be to calculate the probability distribution of the project's present value based on the distribution of each variable and its correlation with other variables.

A conventional cost-benefit analysis however approaches the problem differently. It starts with the expected value for each variable and uses these to calculate an estimate of the project's present value. This estimate will usually be wrong. The reason is simply that the expected value of a function of correlated variables usually does not equal the function of the expected values of the correlated variables.

In defense of conventional cost-benefit analysis one might claim that for many types of projects most variables can safely be assumed to be uncorrelated. This means that the expected present value can be calculated as a function of the variables' expected values.

This reasoning is correct as far as it goes. It ignores however that decision makers are not primarily interested in a project's expected present value. Rather they can be said to maximize expected utility. As investors decision makers will presumably be risk averse. Even decision makers such as government bureaucrats investing other people's money can be quite risk averse in order to avoid blame for failed projects. In some instances a decision maker may adopt a risk loving attitude. This is the case for example when several substitute projects are conducted. They are substitute in the sense that only the best performance will be used such as substitute R & D projects. If the decision maker performs a cost-benefit study of one project without being in a position to judge the total effect of all projects he will nevertheless realize that higher outcomes for his project imply a greater chance that it will beat the other projects. Thus he values higher outcomes disproportionally higher which is equivalent to having a risk-loving utility function.

A decision maker trying to assess the expected utility of a project finds little help in being told a project's expected value. He needs to know the probability distribution.

The traditional answer to outcome uncertainty is to perform sensitivity analyses. These give a good idea of the range of possible outcomes. Unfortunately they convey very little information about how likely different outcomes are.

In the following we confirm these arguments first in a simple and somewhat artificial laboratory experiment. Here subjects are confronted with stylized investment situations and varying degrees of cost-benefit information. Then we report an experiment using real energy-saving projects and real decision makers.

11.2 A laboratory experiment

In this experiment subjects are confronted with a simple investment opportunity that is designed to resemble, in a stylized way, a real situation where costs and benefits must be weighed against each other. The idea is to present investment opportunities that have objectively verifiable expected values that can be compared to subjects' judgements of whether to accept or reject the investments. To achieve this a number of figures are manipulated so that subjects cannot easily calculate the expected value. Instead they must rely on the kind of intuition that people often use to make judgments in complex situations.

There are twelve treatments. These treatments divide into four different investment opportunities. For each type of investment there are three different information levels. One group is only shown the description of the investment (NULL). This group bases its decisions on intuition and whatever calculations can be performed during the experiments duration. The second group is in addition given a conventional cost-benefit study including a sensitivity analysis (CB). The third group does not see the cost-benefit study but only the investment description and a diagram of the density distribution of possible outcomes (DI).

In all treatments subjects are shown a series of 10 uncertain costs and benefits that together comprise the investment. Specifically they are shown a list of 10 figures and are told that each figure multiplied with the outcome of the cast of a die represents a cost or benefit. The cast of a die is a form of uncertainty that people are very familiar with and that subjects, regardless of background, should comprehend easily.

The sum of the 10 costs and benefits yields the outcome of the project. We call this value X. In addition different outcomes have different utilities. Subjects are shown the utility function U and are told that U determines the value of the outcome. They are also shown a graph of the function U to help them acquire an intuitive understanding of the function.

Finally, subjects are asked to judge whether the investment should be undertaken, that is, whether U is positive or negative.

Table 11.1 shows the series of numbers, the utility functions, the cost-benefit information, and the distribution of X. Based on pilot experiments the series of numbers were chosen such that in two cases the cost-benefit analysis was suspected to improve decision making and in two cases it was hypothesized to make decisions less efficient. Investment 1 and 4 have utilities that are of the opposite sign as the expected value of X shown in the cost-benefit information. Thus subjects may be misled by the cost-benefit information.

Table 11.1 Description of the four series of experiments

Series	U	E(U)	Cost-benefit
-1,-1,-1, -1,-1,-2, -2,-3,-4,14	$U=X^2$ if $X>0$ $U=X$ if $X<0$	195	$X=-7$ $X(1)=-2, X(2)=-4$ $X(3)=-6, X(4)=-8$ $X(5)=-10, X(6)=-12$ $X_{max}=68, X_{min}=-82$
-1,-1,-1, -1,-2,-2, -3,-4,8,8	$U=X^2$ if $X>0$ $U=X$ if $X<0$	301	$X=3.5$ $X(1)=1, X(2)=2$ $X(3)=3, X(4)=4$ $X(5)=5, X(6)=6$ $X_{max}=81, X_{min}=-74$
-1,-1,-1, -1,-1,-2, -2,-3,-4,15	$U=X$ if $X>0$ $U=X^2$ if $X<0$	-465	$X=-3.5$ $X(1)=-1, X(2)=-2$ $X(3)=-3, X(4)=-4$ $X(5)=-5, X(6)=-6$ $X_{max}=74, X_{min}=-81$
-1,-1,-1, -1,-2,-2, -3,-4,8,9	$U=X$ if $X>0$ $U=X^2$ if $X<0$	-142	$X=7$ $X(1)=2, X(2)=4$ $X(3)=6, X(4)=8$ $X(5)=10, X(6)=12$ $X_{max}=87, X_{min}=-73$

In the CB treatment subjects were shown the value of X under the assumption that all casts of the die were uniformly 1, 2, 3, 3.5 (expected value), 4, 5, 6. In addition they are shown the largest possible X (X_{max}) and the smallest possible X (X_{min}).

In the DI treatment subjects were shown the probability distribution of X. However they are not shown the cost benefit information.

There were 117 subjects and each subject made a judgement on four different investments. This meant that there were 39 subjects in each of the twelve treatments. For each judgement five minutes were allowed which is not short considering how simple the investments are. Further, subjects were promised a reward of 50 SEK if the four judgements were all correct.

Table 11.2 shows how subjects responded. The cost-benefit analysis improved decision making in two treatments and worsened it in two treatments. The cost benefit studies induced mistakes in investment 1 and 4 where most of the X values shown to subjects had the opposite sign of the true E(U) value. This indicates that subjects take impression of the cost-benefit figures and have difficulty drawing correct conclusions about the actual distribution of X and U values.

Instead showing the distribution in the DI treatment yielded a relatively large improvement in the results in all four investments.

Table 11.2 Correct decisions in percent

	Series				
	1	2	3	4	Total
NULL	62	67	74	59	65
CB	54	78	87	44[a]	65
DI	82[a,b]	80	90	72[b]	88[c]

[a] Differs from NULL at .95 level of significance.
[b] Differs from CB at .95 level of significance.
[c] Differs from NULL at .999 level of significance.

In series 1 and 4 subjects were obviously misled by the cost-benefit information. Since they performed well under the DI treatment the natural interpretation is that the cost-benefit information was unhelpful because it did not allow a good judgement about the probability distribution of outcomes.

We have tried to control the robustness of these results in a number of ways. First we checked whether biases due to learning might have arisen. This was not found to be the case, which is not surprising since the sequence of investments and treatments for each subject was randomized.

Second, subjects were asked a number of control questions designed to test how they approached the problem. Of the 39 subjects 11 had calculated the expected value in the NULL treatment but only 4 had calculated other values. This confirms that the CB treatment added to subjects' information set.

11.3 A real life experiment

Laboratory experiments often lack external validity. Therefore we performed an experiment using practitioners of energy policy. These practitioners evaluate technologically innovative energy projects and decide whether a state subsidy should be granted. Their task is to select projects that are socially valuable but that the applicant would not conduct without the subsidy.

For this experiment we chose a series of projects that build on an existing technology. The purpose of subsidies was to accelerate the diffusion of these technologies by supporting initial installations or demonstration plants.

To improve decision making the Swedish Department of Energy commissioned a computerized cost-benefit model (CCB). This model was designed to calculate not merely expected values but also to take account of all uncertainty in the input variables.

This experiment was performed during the period in which the CCB was introduced. It differs in some aspects from the laboratory experiment. First, practitioners traditionally used a combination of intuition and simple cost-benefit methods. So in this experiment we do not distinguish between these two. Instead there are only two treatments namely evaluation by the traditional method (Trad) and by the CCB method.

Second, in this experiment we do not simply ask for "yes" or "no" judgements. Rather we identify estimates of the projects' private and social values.

Third, the main disadvantage of the real life experiment as compared to the laboratory experiment is that one does not know the true value of the projects. As a result one cannot ascertain which decision method is better in an absolute sense.

Instead we show the following. Project evaluators (PE) usually arrive at similar conclusions about the projects' private return. These corresponded well with the results of the CCB. However PEs arrived at very different conclusions about the projects' social value. The CCB estimates show a much smaller variance. Also the PEs estimates of social value show a bias towards their estimate of the private value. In particular, when the private value was negative PEs were rarely prepared to suggest that the social value could be positive.

In an interview with each PE we tried to pin down exactly how they had estimated the social value. This too indicate that the rules of thumb that were used give rise to significant biases as compared to the CCB.

The projects

The experiment was performed using 6 actual projects. In each case a company or organisation had applied for a subsidy for projects that could demonstrate some extant but unproven electricity-saving technology. If successful each of these technologies could be imitated widely.

The decision making problem in these cases consisted of avoiding two mistakes. First, subsidies should not be given to projects that the applicants would have conducted anyhow. Second, subsidies should not be granted if the social value of demonstrating a technology is smaller than the required subsidy.

Subsidies should thus be granted to projects that are too risky or have a negative expected value to the applicant, but where the social value of demonstrating the technology is large.

An important aspect of these projects is that they can have a positive expected social value even though the expected private value is negative. The reason is that if the technology turns out to be successful then it will be imitated by many, generating a large social value in the process.

The CCB

The CCB captures uncertainty in the input variables. For each variable a user enters a minimum, medium, and maximum value reflecting the uncertainty perceived by the decision maker. This one does for all costs of the project using the new technology and for all costs of the best alternative to this technology.

The CCB then calculates the probability distribution of the difference in costs between the project and the alternative. From this one gets the probability distribution of the net present value of the project.

The CCB then calculates a diffusion curve for the new technology based on empirical experience with similar technologies. The diffusion curve is a function of the project's net present value. If the project incurred a loss no diffusion occurs; otherwise the speed of diffusion is an increasing function of net present value. From the distribution of net present values one has thus derived a probability distribution of diffusion paths.

Finally the CCB calculates the expected social value from the distribution of diffusion paths.

The experimental setting

Five subjects with considerable experience of project evaluation and a good knowledge of the technologies involved were asked to estimate each of the six projects' private and social values. Then they were asked to enter their estimates of input costs in terms of a minimum, medium, and maximum value into the CCB. Finally, in the form of a post-experimental interview, an attempt was made to discover what caused differences in subjects' responses.

There was no explicit time limit during which judgments had to be made.

Results

For all six projects the pattern of results turns out to be quite similar. The CCB and the project evaluators (PE) come to rather similar conclusions about the private expected value of the projects. Their judgements about the projects' social value diverges considerably. The latter effect is especially pronounced for projects that are

Figure 11.1.

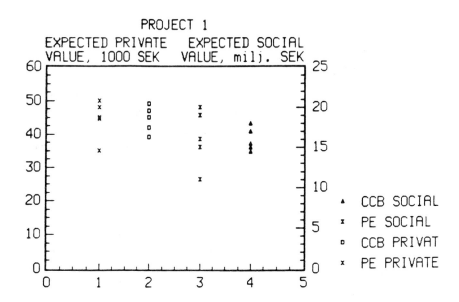

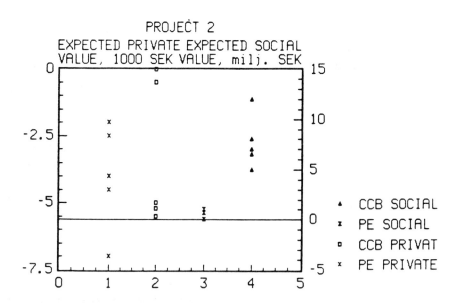

Figure 11.1.

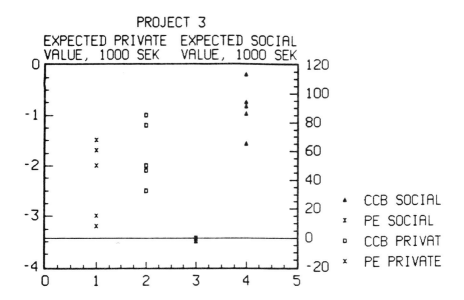

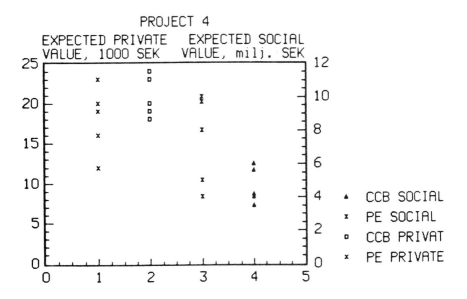

Figure 11.1.

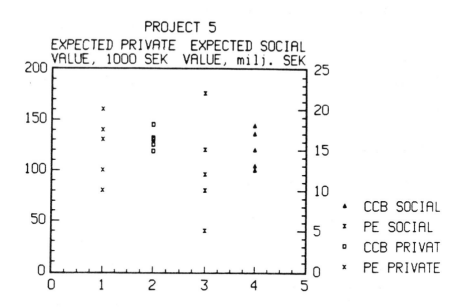

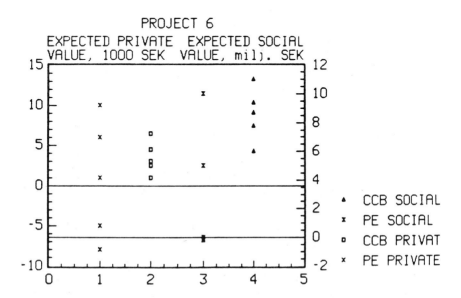

expected to have negative private values. Also, compared to the CCB, the project evaluators exhibit a considerable bias to give the social value the same sign as the private value.

Figure 11.1 shows the private and social values estimated by project evaluators.

For all projects one can observe a wide divergence in the social value that PEs estimate. This divergence is usually larger than the CCB estimate. This indicates that part of the divergence in PE estimates is due to difficulty in calculating the social value rather than uncertainty about input values.

Project 2 and 3 are estimated to have negative private expected values. For these projects there is a drastic difference between PE and CCB estimates of social values. PEs do not concede the possibility that the project turns out better than expected and in that case generates a large social externality. Project 6 involves a very uncertain technology. PEs were divided about the expected private value of projects. Characteristically those that believed in a negative expected private value did not believe that projects had any social value either.

The post-experimental interviews indicate that differences in PEs' estimates of cost items were of minor importance. Rather PEs had a mistrust and poor understanding of the argument for why projects with negative expected private values could still have positive expected social values. They also had considerable difficulties in conceptualizing the consequences of the spread in possible private value outcomes. Instead their estimates of the social value were often based on simple calculations such as the private expected value multiplied with a fraction of the potential market.

After an explanation of how the CCB operates subjects usually agreed with the probability distribution of possible private values generated by the CCB. However they considered any calculation of social values to be extremely uncertain and questioned how robust the results of the CCB were.

Conclusion

Both the laboratory- and the real life experiments indicate that cost-benefit studies of the sort normally used in practice have considerable shortcomings and may in some instances be detrimental to efficient decisionmaking. In particular there seems to be a bias on the part of decision makers to give excessive weight to cost-benefit results even when it is known that other factors, not addressed by the cost-benefit, can considerably alter the result.

Therefore it seems to be worthwhile to design more complicated cost-benefit methods that provide a better idea of the likelihood of different outcomes.

12 The incentive subsidy

12.1 The impossibility of incentive compatibility

One strand of economic literature has analysed procurement contracts in situations where the government intends to purchase a product from a firm. With uncertainty about costs the problem there is how to design an incentive compatible contract that induces cost minimization (Weitzman, 1980) and leads firms to present honest initial cost estimates.

The problem set in this chapter differs from that of the procurement literature in three ways. First, one wants the firm to reveal not only cost estimates, but also estimates of the value of the invention, since in the case of a subsidy the government is not a buyer who is in a good position to value the prospective invention himself. Second, the firm should refrain from misrepresenting its project not just when the project is socially worthless, but also when it is privately profitable. Third, the firm should have an incentive to maximize the value of the prospective invention in addition to minimizing costs.

This chapter begins by showing why granting a general subsidy may not be an efficient way of correcting R & D externalities. This justifies the search for a subsidy mechanism that helps to select projects that should be subsidized. Then it is shown that an incentive compatible policy that accomplishes this selection perfectly may not be attainable. Finally, a subsidy mechanism is examined that is incentive compatible in many circumstances and, in addition, has a very simple structure. This mechanism we call the "incentive subsidy".

First best policy

The general argument for subsidizing privately conducted R & D is that firms cannot always appropriate the entire social return of an invention to themselves; the invention has a positive externality. As a result firms do not always research enough from a social point of view without subsidies.

The traditionally advocated solution to positive externalities is to reimburse firms for the difference between the private and the social value of the activity generating the externality, in this case for the inventions that are made. Call this the "first-best" solution.

In a world with convex production functions every single invention must be subsidized that displays a difference between its private and its social value. In contrast, when there are non-convex research production functions giving rise to discrete research projects the government can save public funds by subsidizing only those projects that firms would not have conducted without the subsidy.[25]

To show this suppose there exists a set of discrete project ideas 1..n with fixed private returns to inventing $V_1..V_n$. The social values of these projects are larger, ranging from $V_1(1 + s) .. V_n(1 + s)$ where $S>0$. The projects have costs c1..cn.

Of the n projects m have a negative private value, m<n. All n projects have a positive social value. The shadow price of subsidy funds is r. Then the total social value of all projects conducted when all inventions are subsidized in the way suggested by the "first-best" solution is:

(9) $SV = (V_x(1 + s) - c_x - rsV_x)$

If only projects with a negative private value are subsidized then the total social value is:

(10) $SV = (Vx(1 + s) - c_x) - rsV_x$

The total social value in (10) is larger than in (9) because m<n. Actually the social value in (9) could be increased even further because not the entire s V_x needs to be paid for each project to get firms to research the project.

Incentive compatibility

If a government agency distributing subsidies can observe perfectly which projects firms would conduct even without a subsidy then the efficiency gain mentioned above could be easily exploited. Without perfect information there are two paths to take. Either the government can try to guess which projects the firm would conduct anyway and accept mistakes that occur; or one can try to devise the subsidy in such a way that firms do not even apply with projects they would conduct anyhow.

Pursuing the latter approach we define an incentive compatible subsidy as one that fulfills the following requirements:

1. The firm should not apply with any project it would conduct anyhow. For a risk neutral firm this means that for no project for which the private expected profit without a subsidy, $E(R^u)$, is greater than zero should the expected profit with a subsidy, $E(R^s)$, exceed $E(R^u)$.

2. If $E(R^u)$ is below zero then $E(R^s)$ should be smaller than zero only if the expected social value Se is smaller than zero.

---

[25] That projects are discrete does not mean that it is impossible to conduct them with other amounts of inputs but only that it is not economical to do so. For example a project may have increasing returns to scale 1up to a point and sharply falling returns to scale after that point. Then that point probably defines how the projects should be conducted regardless of the cost of inputs or the value of the output.

3. If subsidized the firm should expect a higher profit $E(R^s)$ for higher social values S.

The proof we are about to present makes the point that an incentive compatible policy that is applicable to very general circumstances is impossible. The proof proceeds by showing that with assumptions reflecting plausible circumstances one can find two R & D projects such that any policy that is incentive compatible for one project is not incentive compatible for the other.

The first assumption is that the subsidizing agency knows nothing about a research project in advance except what it is told by the firm. After the project has been conducted the subsidizing agency learns the actual value of the project and its actual cost, in effect it learns S and R, the social and private values of the project. Since the subsidizing agency does not learn about the projects until afterwards this also means that a subsidy policy must be incentive compatible for any possible project that firms may apply with.

The second assumption is that firms are risk averse. The third assumption is that firms, when they are indifferent between two alternatives, do not necessarily choose the one with a higher social value.

Proof

$E(R^u)$ is the profit the firm expects from a project if it is not subsidized. $E(R^s)$ is the profit the firm expects if it is subsidized. S is the ex post social value of the project, and S^e is the expected social value. Then an incentive compatible policy must meet the following three requirements.

1. If $R^u > 0$ or $S^e < 0$ then $R^s < 0$.

2. If $S^e > 0$ and $R^u < 0$ then $R^s > 0$.

3. $dR^s/dS > 0$.

Assume that a firm can apply with two projects. Both projects have a positive social value and should therefore be conducted. Project A yields $R^u = 1$ with absolute certainty. It should therefore not be subsidized. Project B has a negative expected private profit, but there is a chance of earning a $R^u = 1$. An incentive compatible policy for project A must then ensure that for $R^u = 1$, $R^s < 0$. For project B if $R^u = 1$, $R^s > 0$. Thus R^s can not be set in a way that is incentive compatible for both projects.

A more detailed analysis of this proof is given in Fölster (1988b). This proof depends on a number of assumptions that can be relaxed. For example, if the policy maker has some ex ante information about the projects an incentive compatible policy can probably be designed for some ranges of projects. The next section proposes a policy that is incentive compatible for most projects out of a plausible range of projects that firms can be expected to apply with.

12.2 The incentive subsidy

In the previous section it was shown that a general incentive compatible policy suitable to all projects firms might apply with cannot be devised. Here the incentive subsidy is suggested as one of the most promising second-best alternatives.

First, we show how the incentive subsidy works and why it comes close to achieving incentive compatibility under reasonable circumstances. Then the incentive subsidy is shown to be more effective than normal subsidies and conditional loans. This result is then comfirmed in a simulation, showing for a hypothetical distribution of projects that the policy proposed here performs increasingly better than the conditional loan or project specific grants when the government's information about projects deteriorates.

The incentive subsidy eliminates the need for an exante judgement by the government agency on whether a project should be subsidized. Instead the exact size of the subsidy is determined after the project has been conducted. This ex post adjustment of the subsidy is done in such a way that the firm usually applies for the subsidy only when it should be subsidized from a social point of view. Under the incentive subsidy firms are reimbursed for any private loss they make and any private profit is taxed away; in addition the firm receives a small fraction of the invention's social value. As a result it will conduct a subsidized project in a way that maximizes social value. Also it applies only if its project has a positive expected social value and a small or negative expected private value.

A possible objection to the incentive subsidy is that it requires estimation of research projects' social and private value. Such estimates can be extremely uncertain.[26] This uncertainty however is not a serious problem for the incentive subsidy. It is shown that even large errors in the estimates of social value affect the efficiency of the incentive subsidy rather little. The reason is that the firm will not know in which direction the government errs until after the project is completed. Also, in comparison to the other subsidy forms an error is much less serious because the estimate is made ex post with the results in hand rather than ex ante as required by the normal subsidy

---

[26] The claimis sometimes made that its virtually impossible to value many inventions. As a counter argument one need look no further than the stock market where venture capital firms with risky research projects are valued by private agents all the time. So the real question is not whether these values can be estimated, but rather how seriously mistakes in this valuation damage the efficiency of the policy.

and the conditional loan. More about the estimation of social and private values is given below.

Under the incentive subsidy scheme firms must apply prior to the commencement of a project. At that time firms may or may not receive an advance loan.[27] The important thing is that the exact size of the subsidy is not determined until after the projects has been completed.

The incentive subsidy contains a component that compensates the firm for a loss or taxes away a gain it makes on the project. In addition the firm is rewarded a fraction a of the social value S.[28] This induces socially efficient research. The subsidy g is then as follows, where R is the private return, and the tax of profit or compensation for loss corresponds to $-R$:

$$g = -R + aS.$$

The expected value of researching to the firm with the subsidy is $R^s = E(R + g) = aS^c$. Here S^c is the expected social value. As a result the firm does not apply with any project that has a negative expected social value.

Since the firm is rewarded for maximizing the social value it conducts the project efficiently, minimizing costs and maximizing the social value of the innovation.[29]

When a project has positive private return, so that $R > 0$, then the firm usually looses by applying to the subsidy system because the private return will be taxed away. However, there is a special case, as mentioned above, where the incentive subsidy is not perfectly incentive compatible. The firm will lie about some projects it would have researched even without the subsidy, and will receive funding for them. If the firm is risk neutral this occurs for projects that have an expected unsubsidized return R^u:

$$0 < R^u < aS^c.$$

---

[27] Advance loans become necessary only when capital markets do not function perfectly. This may be the case in practice. Correcting imperfections in the capital market should be treated as a separate problem however, requiring a separete remedy. The incentive subsidy as such solves only one market failure. Amending the incentive subsidy with loans ameliorates a different market failure and is therefore not further considered here.

[28] The social value can be calculated by following a set of rules of thumb. The firm may know these rules in advance, but it will not know how the government judges specific values until the project has been concluded. In practice it may be debatable when exactly a project is concluded. It is hard to believe however that this constitutes a major problem.

[29] Since the incentive subsidy rewards a firm for increases in social value it may also be used to increase the rate of diffusion of a technology. For example if the firm can show that it has helped other firms to use its invention as well then the estimated social value will be greater and the firm will earn a greater return.

As long as the government correctly estimates R and S after the research has been conducted, a can be held extremely low, provided only that the firm does not treat it as negligible. Then there are probably only a few projects within any reasonable distribution for which the incentive subsidy fails.

If firms are risk averse the incentive subsidy also acts as an insurance. Suppose a firm has a project with a positive R^u that is too risky for it to conduct. Then without the subsidy it gains nothing, but with the subsidy it expects a small return aS^e involving little risk. So it will opt for the subsidy. The government can then expect a net return of $R^u - aS^e$. This is akin to an insurance where the premium is paid afterwards.

Determining the optimal level of a when government estimates contain an error

If a firm is not risk averse then even a small value for the parameter a will induce it to research in a socially optimal way. Things are slightly more complicated when firms are risk averse and the government makes ex post mistakes in determining the value of R and S. Suppose first that there are no systematic mistakes, so that the firm expects the goverment to be correct on average. Then joining the incentive subsidy will become more of a risky business for firms. To compensate for this the level of a must be set at a somewhat higher level as shown below. The important point is however that even ex post government mistakes in judging R and S probably do not affect the efficiency of the incentive subsidy greatly as long as the mistakes are not systematic and predictable by firms.

To show what the optimal level of a is for a given project, suppose that the government forms ex post estimates of the social and private values of a project, each containing the error, e_R and e_S respectively, with zero means and any standard deviation:

$$S^g = S + e_S \qquad R^g = R + e_R.$$

Both R and S are known to the firm and are assumed to be functions of a firm effort w, so that $R = R(w)$ and $S = S(w)$. It is assumed that $S > R$ and that both are convex differentiable functions of w with $S'(w) > 0$, $R'(w) > 0$, $S''(w) < 0$ and $R''(w) < 0$.[30] It follows that the socially optimal w^S is larger than or equal to the privately w^R. Further it is assumed that there non-convexities in the industry research production set. This means that some research projects may be conducted in a socially optimal way even without a subsidy. If this

---

[30] Empirical studies tend to find that social returns to inventions are much larger than private returns.

were not the case then the best policy could be merely to reimburse all forms for the difference between social and private values. The non-convexity however means that the government may save public funds by selectively subsidizing only projects that firms would not conduct otherwise.

With the incentive subsidy the firm expects a return of

$$V = R + g = aS^g - R^g + R$$

and it maximizes a utility function assumed to take the following simple form: $U = E(V) - mo_V$. Then o_V, the standard deviation of the firm's return is

$$o_V = E(aS^g - R^g + R - aS)^2 = (ae_S - e_R)^2.$$

This shows that the standard deviation of V is independent of w. So the firm maximizes its utility by setting $U'(w) = 0$. This yields the result that the firm sets w to the socially optimal value at w^S, which is also the w at which $S'(w) = 0$.

The government in turn maximizes

$E(S-rV)$ s.t. $U > 0$ and $a \geq 0$.

Here r is the opportunity cost of raising public funds. The constraints exist to ensure that the firm will rearch under the subsidy scheme and to ensure that it maximizes social value. Taking the derivative shows that the parameter a is then set as small as possible to just fulfill the constraints:

$a > (mo_V)/S$ and $a \geq 0$.

This shows that as long as the government makes no systematic error, so that the error's expected value is zero, firms will set w to its socially optimal level regardless of the choice of a – provided that the constraints are satisfied.

Of course the government will not know the level of risk aversion among firms, so it may have to set a common a for all firms. The less accurate a is set then the larger the chance of not fulfilling the constraint $U > 0$ exactly with as small an a as possible. This implies that some errors are committed with the incentive subsidy.

Things become worse when the government makes systematic mistakes. Suppose as an extreme case of neglect, it never takes firm effort w into account when reimbursing the firm. Then $V = aS^g - R^g + R - w$. The optimal w for the firm is then where

$$U'(w) = aS'(w) - 1 = 0.$$

This means that w is set at a level below the socially optimal level. Further, the firm increases w as a increases and it reaches its socially optimal value only when $a = 1$.

This means that if systematic mistakes become unavoidable, say in the case where a single inventor is subsidized whose effort cannot be observed, then the problem is transformed into a traditional principal agent problem. In this case the incentive subsidy requires a larger a; but a large a implies a wider range of projects where firms cheat and apply with projects they would have conducted anyhow. While the efficiency of the incentive subsidy is impaired when the government commits systematic errors the other two subsidy forms suffer detrimental effects that are at least as large. This is shown in the following sections. The reason is that the systematic error also leads to mistakes in granting normal subsidies or conditional loans.

The incentive subsidy in comparison

This section presents the theoretical arguments that support the incentive subsidy as a superior alternative to normal subsidies and conditional loans.

The arguments are based on the following assumptions. The government can estimate the social value of a research project before (ex ante) it is conducted and afterwards (ex post). The ex post evaluation is always as least as accurate as the ex ante evaluation, but often much more accurate.

The first principle is that a subsidy is more effective if the decision to subsidize is based on more accurate information. This shows why the incentive subsidy and the conditional loan outperform the normal subsidy. With a normal subsidy the government evaluates a project ex ante. Then it signs a cheque with few strings attached. Information that emerges ex post − but that the firm may have secretly known all along − is ignored.

The conditional loan is more refined.[31] Here the firms required to pay back its subsidy if the project returns a private profit. The government can always set the size of the conditional loan exactly equal to the normal subsidy and, neglecting the available ex post information, grant this loan to exactly the same firms that would have received the normal subsidy. Neglecting all ex post information means that the loan is never retrieved. It follows that one can always do at least as well with the conditional loan as with the normal subsidy policy.

Since the government uses the ex post information, available under the conditional loan scheme, only when this is expected to raise social value, the conditional loan will always be a better policy tool when the ex post information is better than ex ante information.

---

[31] For example STU, the main government agency dispensing research subsidies in Sweden, grants a considerable fraction of its budget in the form of conditional loans. Of these subsidies roughly 25 % are repaid.

Similarly the incentive subsidy can be made to grant exactly the same sums to firms as the normal subsidy by neglecting ex post information and setting the parameter *a* to zero.

The normal subsidy has two further problems apart from using ex ante information. First, it does not reward increases in social value. Second, it does not reduce the risk to firms as much as the conditional loan and the incentive subsidy. Both of the latter pay out larger sums when the project fails than when it succeeds. Since a risk averse firm values a unit subsidy more in the event that it is making a loss than when it is making a profit the same expected value of a subsidy raises utility less with the normal subsidy. This also means that one can get the firm to research, by raising its expected utility above zero, with a lower level of expected government handouts under the conditional loan and incentive subsidy. Since government handouts have an opportunity cost it follows that a lower government expenditure is a definite advantage.[32]

Comparing the conditional loan with the incentive subsidy is slightly more complicated. The main problem with the conditional loan is that one cannot tax the firm if the project turns out to be privately profitable. As a result firms will apply for the loan even with projects that they would conduct anyhow, but that have a chance of returning a private loss. Another problem is the fact that the conditional loan does not reward improvements in social value.

The incentive subsidy can always be made to perform at least as well as the conditional loan. This is apparent from the fact that the exact size of the incentive subsidy can be adjusted to any desired amount based on all available ex post information about the private and social return. When granting a conditional loan on the other hand the size of the potential subsidy must be determined based only on ex ante information. Ex post information can be used only in a very restricted way to determine how much of the loan should be repaid. One can never ask the firm to repay more than it received in the first place. This means that the incentive subsidy can be set at exactly the same level as the conditional loan if the government gives up some of its freedom to act upon ex post information. Assuming that the government only uses the greater freedom with the incentive subsidy when this is expected to raise social value, it follows that the incentive subsidy is better.

More precisely, the incentive subsidy has the following advantages:

1. The first problem with the conditional loan is that it does not reward social efficiency. Thus if the social value of a project can be

---

[32] Public funds have a higher opportunity cost than the firm's funds because they consist of the private opportunity costs of whoever they were taxed from as well as the deadweight loss of taxation.

raised by incurring some extra expenditure then the firm with the conditional loan will never do so, while under the incentive subsidy the government can adjust the parameter *a* to induce the firm to do what is socially efficient.

2. With the conditional loan firms will try to get loans for projects that they would conduct anyhow but that have a chance of failing. The poorer the government's ex ante information is the poorer it will be at weeding out those projects. With the incentive subsidy this type of mistake occurs only for projects where the expected utility of aS is larger than the expected private profit. This ought to be an unusual case since *a* can be set at low level.

3. The conditional loan reduces risk for the firm less than the incentive subsidy because the size of the loan does not vary with the extent of private loss. This menas that somewhat larger payment may be required in order to get the firm to research.

It must be emphasized that this comparison of subsidy policies is valid even if the government makes mistakes in estimation the social value. The reason is that mistakes in estimation the social value affect all policies. While systematic mistakes have similar effects for all policies, random errors are less serious for the incentive subsidy because they are committed after the firm has conducted its project. Since the firm does not know in which direction the error will occur it will presumably research in the socially most efficient way.

A simulation

The comparison of subsidy policies in the previous sections has isolated the factors that determine the relative efficiency of the policies without really shedding much light on the quantitative importance of the efficiency differences.

This is a difficult theoretical task mainly due to the problems in specifying general optimality conditions for the size of subsidies over a distribution of distinct projects when adverse selection and cheating must taken into account.

Instead this problem is solved numerically in a simulation model. The simulation has been performed a large number of times with varying assumptions. The pattern of results is always similar. Here a typical set of results is presented. It is shown that the incentive subsidy suggested in this paper performs better than the conditional loan and the normal subsidy. However, when the government has perfect information the difference between the subsidy policies is small. When the government has poor information the conditional loan and the normal subsidy perform considerably worse than the incentive subsidy.

The simulation is performed over a range of 30 projects. For each type of subsidy policy the simulation model determines whether and how the project is conducted by firms and what the social value is. The social values are then added to show the efficiency of a policy over the entire range of projects. The detailed assumptions of the model are supplied in the appendix. In short, each project contains an uncertainty of succeeding better or worse. Firms calculate what subsidy they are to receive under each possible project outcome and thus arrive at an expected private value and a utility level (to account for risk aversion). Of the 30 projects 8 have negative social and private values, 11 have a positive social value and negative private utility level, and 11 have positive private and social values.

Table 12.1 Percentage increase in social value over the non-subsidized outcome

	Perfect information	Small government error	Large government error	Systematic government error
1. Incentive subsidy	26	23	16	9
2. Normal subsidy	19	12	−6	2
3. Conditional loan	22	17	5	5
4. Hypothetical perfectly incentive compatible subsidy	28	25	19	12

Table 12.1 shows a typical set of results. The values shown are percentage increases in social value due to the respective subsidy policy being introduced. Apart from the three subsidy policies discussed in this paper the table also shows results for a hypothetical perfectly incentive compatible policy. This represents the maximum increase in social value possible, in effect when firms act as if their interests were identical with the government's.

Four different assumptions are made about the accuracy of the government's estimates of social values. The first column assumes no errors at all. The second and third column assume a small and a severe random error. The fourth column assumes a systematic overvaluation of the true social values. The specific representation of these errors is explained in the appendix.

The results show that when the government is well informed all subsidy policies perform relatively well. When the government is not well informed then the normal subsidy and the conditional loan perform relatively worse while the incentive policy still performs quite well.

When there is a systematic bias in the government's evaluation of social values then all policies perform worse, but the incentive subsidy retains its relative advantage.

Conclusion

It is argued that the incentive subsidy is a better policy than either the normal subsidy policy or the conditional loan that are commonly used in many countries.

Theoretical arguments lead to the conclusion that the conditional loan is a better policy than normal project grants and that the incentive subsidy is a better policy than the conditional loan.

Finally, a simulation of the different policies over a range of hypothetical projects compares the policies when the government has imperfect information about the projects. It is shown that the worse the government's information is the better the incentive subsidy performs relative to other policies.

Appendix to chapter 12

All firms have the same utility function with constant absolute risk aversion. Due to the risk aversion not all projects with positive expected private values have positive expected utilities.

It is assumed that public funds have an opportunity cost of 10%. The projects themselves have a value that contains a constant component T, and a component $t \ln(w)$ that the firms determines itself by choosing an effort w. In addition there is a random component o that has a 50% chance of being added or subtracted. The expected social value of a project is then:

(A1) $S^e = T + t \ln(w+1)(1+s) - w + 0.5o - 0.5o.$

The social value of a project is higher than its private value, due to the parameter s, that is set equal to 0.7 here. So the private expected value is:

(A2) $R^u = T + t \ln(w+1) - w + 0.5o - 0.5o.$

Maximizing with respect to w gives an optimal private $w^P = t - 1$ and an optimal social $w_s = t(1+s) - 1$. In the simulation T increases in increments of 1 from -15 to 14 thus creating 30 projects. t is set to 4 and o to 10.

To account for risk aversion the form for constant absolute risk aversion is used: $U = 1/q(1 - \exp[-qX])$. q is set to 0.13 and X is the actual firm return.

With perfect government information the subsidies are calculated as follows:

1. Hypothetical perfectly incentive compatible subsidy: this is the amount required to compensate firms for researching in a socially optimal way, assuming that there are no incentive problems. Thus

if R^u is negative then $g = -R^u + (w_S - w_P)$ and if R^u is positive then $g = w_S - w_P$.
2. Incentive subsidy: the parameter a is set to 5%
3. Normal subsidy: for all projects that have $S^e - rg > 0$ the subsidy is set so that the firm will just research, $EU = 0$.
4. Conditional loan: as for the normal subsidy, given that the firm has to repay if $R > 0$.

When the government does not have perfect information, then it makes mistakes in estimating the project parameter o. The error e is assumed to follow a binary distribution so that o is estimated at $(o+e)$ or $(o-e)$, each with a 50% chance. e is set at the levels 3 and 8. The policies are then set as follows:

1. Incentive subsidy: the private return and the social value are estimated with an error. The optimal policy is just as in the perfect information case.
2. Normal subsidy: the social and private values are estimated with an error, leading to mistakes in deciding what the level of subsidy should be. The optimal subsidy turns out to be 0.6 times the perfect information subsidy when $e=3$, and 0 when $e=8$.
3. Conditional loan: the social and private values are estimated with an error, leading to mistakes in deciding what the level of the loan should be and how much be repaid. The optimal loan turns out to be 0.8 times the perfect information loan when $e=3$, and 0.7 times the perfect information case when $e=8$.

When the government commits systematic errors, e.g. consistently overestimation the social value, the subsidies are calculated as in the perfect information case above. The only difference is that now the government's estimate of social value is taken to be twice the true social value.

Bibliography

Anderson, N. H. Looking for configurality in clinical judgements. *Psychological bulletin,* 1972, 72, 93–102.

Andersson, R. *Teknikbedömning på energiområdet.* Stockholm: Byggforskningsrådet, 1986.

Allen, T., Otterback, J. M., Sirbu, M. A., Ashford, N. A., and Holloman, J. H. Government influence on the process of innovation in Europe and Japan. *Research policy,* 1978, 7, 124–149.

Allen, T., Hyman, D., and Pinckney, D. Transferring technology to the small manufacturing firm: A study of technology transfer in three countries. *Research policy* 1983, 12, 199–211.

Arrow, K. Economic welfare and the allocation of resources for inventions. In R. R. Nelson (ed.), *The rate and direction of inventive activity.* Princeton: Princeton University Press, 1962.

Basi, B. A., Carey, K. J., and Twark, R. D. Comparison of the accuracy of corporate and security analysts' forecasts of earnings. *Accounting review,* 1976, 51, 244–254.

Bernstein, J. I. *Research and development, tax incentives and the structure of production and financing.* University of Toronto Press, Toronto, 1987.

Bernstein, J. I. The structure of Canadian inter-industry R & D spillovers, and the rates of return to R & D. *The journal of industrial economics,* 1989, 37, 315–328.

Bertin, G. Y. and Wyatt, S. *Multinationals & industrial property.* Harvester-Wheatsheaf: England, 1988.

Bohm, P. and Lind, H. Sysselsättningseffekter av sänkt arbetsgivaravgift i Norrbotten 1984–1986. Research papers in Economics 1988:1 RS, Stockholm University.

Bozeman, B. and Link, A. Tax incentives for R & D: A critical evaluation. *Research policy,* 1984, 13, 21- 31.

Brockhoff, K. The measurement of goal attainment of governmental R & D support. *Research policy,* 1983, 12, 171–182.

Carmichael, J. The effects of mission-oriented public R & D spending on private industry. *Journal of finance,* 1981, 36, 617–27.

Copeland, R. M. and Marioni, R. J. Executives' forecasts of earnings per share versus forecasts of naive models. *Journal of business,* 1972, 45, 497–512.

Dasgupta, P. and Maskin, E. The simple economics of research portfolios. *Economic journal,* 97, 1987, 581- 595.

Dasgupta, P. and Stiglitz, J. Uncertainty, industrial structure, and the speed of R & D. *Bell journal of economics,* 1980, 266–293.

Dasgupta P. and Stoneman P. *Economic policy and technologica performance,* Cambridge University Press, Cambridge: 1988.

Dawes, R. M. and Corrigan, B. Linear models in decision making. *Psychological bulletin*, 1974, 81, 95–106.

Dosi, G. Sources, Procedures and microeconomic effects of innovation. *Journal of economic literature*, 1988.

Ergas, H. The importance of technology policy. In Dasgupta, P. and Stoneman, P. (eds.) *Economic policy and technological performance*, Cambridge: Cambridge University Press, 1987.

Ettlie, J. The commercialization of federally sponsored technological innovations. *Research policy*, 1982, 11, 173–192.

Ettlie, J. Policy implications of the innovation process in the U.S. food sector. *Research policy*, 1983, 12, 239–267.

Folmer, H. and Nijkamp, P. Investment premiums: Expensive but hardly effective, *Kyklos*, 1987, 43–72.

Fölster, S. Government coordination of competitive R & D projects: A socially optimal policy using a search-theoretic approach. Unpublished D.Phil. thesis, Oxford, 1986.

Fölster, S. Contracts for R & D subsidies: A comparison of incentive structures. Discussion paper, University of Stockholm, 1987.

Fölster, S. The incentive subsidy for government support of private R & D. *Research policy*, 1988, 17, 105–112.

Fölster, S. Firms' choice of R & D intensity in the presence of aggregate increasing returns to scale. Discussion paper, Industrial Institute for Economic and Social Research, Stockholm, 1988.

Fölster, S. The design of government schemes to stimulate industrial innovation. Discussion paper, The Industrial Institute for Economic and Social Research, Stockholm, 1988.

Goldberg, L. R. Five models of clinical judgement: An empirical comparison between linear and nonlinear representations of the human inference process. *Organizational behavior and human performance*, 1971, 6, 458–479.

Griliches, Z. Hybrid corn: An exploration of the economics of technological change, *Econometrica*, 1957, vol 25, p. 501–22.

Griliches, Z. Returns to research and development expenditures in the private sector. In Kendrick J. W. and Vaccara, B. N. (eds.) *New developments in productivity measurement and analysis.* Chicago: University of Chicago Press, 1980.

Griliches, Z. and Pakes, A. The value of patents as indicators of inventive activity. In Dasgupta, P. and Stoneman, P. (eds.) *Economic policy and technological performance.* Cambridge: Cambridge University Press, 1987.

Gronhaug, K. and Frederiksen, T. Governmental innovation support in Norway. *Research policy*, 1984, 13, 165–173.

Hansson, I. Marginal cost of public funds for different tax instruments and government expenditures. *Scandinavian journal of economics*, 1984, 115–130.

Hall, P. *Technology, innovation & economic policy,* Phillip Allan Publishers, Oxford: 1986.

Hirshleifer, J. Investment decisions under uncertainty: Applications of the state preference approach. *Quarterly journal of economics,* 1966, 80, 252–77.

Holemans, B. and Sleuwaegen, L. Innovation expenditures and the role of government in Belgium. *Research policy,* 17, 1988, 375–379.

Johnson, T. and Kaplan, R. *Relevance lost: The rise and fall of management accounting.* Harvard Business School Press, 1987.

Kamien, M. and Schwartz, N. *Market structure and innovation,* Cambridge: Cambridge University Press, 1982.

Leff, N. H. The Use of policy-science tools in public-sector decision making: Social benefit-cost analysis in the World Bank. *Kyklos,* 1985, 38, 60–76.

Leff, N. H. Policy research for improved organizational performance. *Journal of economic behavior and organization,* 1988, 9, 393–403.

Leonard-Barton, D. Interpersonal communication patters among Swedish and Boston-area entrepeneurs. *Research policy,* 1984, 13, 101–114.

Levin, R. C. A new look at the patent system. *American economic review,* 1986, 76, 199–202.

Levin, R., Klevorick, A., Nelson, R., and Winter, S. Appropriating returns from industrial research and development. *Brookings papers on economic activity,* 1987, 3, 783–820.

Levy, D. M. and Terleckyi, N. E. Government financed R & D and productivity growth: Macroeconomic evidence. National Planning Association. Mimeo, 1981.

Levy, D. M. and Terleckyi, N. E. Effects of government R & D on private R & D investment and productivity: A macro-economic analysis. *Bell journal of economics,* 1983, 551–61.

Lichtenberg, F. R. The relationship between federal contract R & D and company R & D. *American economic review,* 1984, 74(2), 73–78.

Link, A. Basic research and productivity increase in manufacturing: Additional evidence. *American economic review,* 1981, 71(5), 1111–1112.

Little, I. M. D. and Mirrlees, J. Manual of industrial project analysis in developing countries, vol. 11, Paris: *OECD,* 1968.

Loury, G. C. Market structure and innovation. *Quarterly journal of economics,* 1979, 93, 395–410.

MacDonald, S. Theoretically sound: Practically useless? Government grants for industrial R & D in Australia. *Research policy,* 1986, 15, 269–283.

Mansfield, E. R & D and innovation: Some empirical findings. In Griliches (ed.) *R & D, Patents, and productivity*, 1984, Chicago: University of Chicago Press, 1984.

Mansfield, E. How rapidly does new industrial technology leak out? *Journal of industrial economics*, 1985, 217- 233.

Mansfield, E. The R & D tax credit and other technology policy issues. *American economic review*, 1986, 76, 190–194.

Mansfield, E., Rapoport, J., Romeo, A., Wagner, S., and Beardsley, G. Social and private returns from industrial innovation. *Quarterly journal of economics*, 1977.

Mansfield, E., Schwartz, M., and Wagner, S. Imitation costs and patents: An empirical study. *Economic journal*, 1981, 91, 907–18.

Mansfield, E. and Switzer, L. The effects of R & D tax credits and allowances in Canada. *Research policy*, 1985, 14, 97–107.

Meyer-Krahmer, F., Gielow, G., and Kuntze, U. Impacts of government incentives towards industrial innovation. *Research policy*, 1983, 12, 153–169.

Nelson, R. R. (ed.) *Government and technical progress.* Pergamon, New York, 1982.

OECD, *OECD science and technology indicators*, Paris: 1986.

Office of Technology assessment, *Research funding as an investment: Can we measure the returns? – A technical memorandum*, Washington, DC: U.S. Congress, Office of technology assessment, OTA-TM-SETET-36, 1986.

Oshima, K. Technological innovation and industrial research in Japan. *Research policy*, 1984, 13, 285–301.

Pavitt, K. Government policies towards innovation: A review of empirical findings. *Omega*, 1976, 4, 539–558.

Pavitt, K. and Walker, W. Government policies towards industrial innovation: A review. *Research policy*, 1976, 5, 11–97.

Peck, M. Joint R & D: The case of microelectronics and computer technology corporation. *Research policy*, 1986, 15, 219–231.

Robbins, M. D. and Milliken, G. J. Government policies for technological innovation: Criteria for an experimental approach. *Research policy*, 1977, 6, 214–240.

Roessner, D. The local market as a stimulus to industrial innovation. *Research policy*, 1977, 8, 340–362.

Roessner, D. Commercializing solar technology: The government role. *Research policy*, 1984, 13, 235–246.

Rothwell, R. Venture finance, small firms and public policy in the UK. *Research policy*, 14, 1985, 253–265.

Rothwell, R. and Townsend, J. The communication problem of small firms. *Research and development management*, 1973, 3, 151–153.

Rothwell, R. and Zegveld, W. *Industrial innovation and public policy*, Frances Pinter, 1981.

Rubenstein, A. H., Douds, C. F., Geschka, H., Kawase, T., Miller, J. P., Saintpaul, R., and Watkins, D. Management perceptions of government incentives to technological innovation in England, France, West-Germany, and Japan. *Research policy*, 1977, 6, 324–357.

Schnee, J. E. Government programs and the growth of high-technology industries. *Research policy*, 1978, 7, 2–24.

Scott, John T. Firm versus Industry variability in R & D intensity. In Griliches, Z. (ed.) *R & D, Patents, and productivity*. Chicago: The University of Chicago Press, 1984.

Souder, W. and Mandakovic, T. R & D project selection models. *Research management*, 1986, 36–42.

Stiglitz, J. & Weiss, A. Credit rationing in markets with imperfect information. *American Economic Review*, 1981, 71, 393–410.

Stoneman, P. *The economic analysis of technical change*. Oxford University Press, 1983.

Stoneman, P. *The economic analysis of technology policy*, Clarendon Press, Oxford: 1987.

Styrelsen för Teknisk Utveckling. *STU perspektiv 1983*. Stockholm, 1983.

Tassey, G. The technology policy experiment as a policy research tool. *Research policy*, 1985, 14, 39–52.

Taylor, C. T. and Silberston, Z. A. *The economic impact of the patent system: A study of the british experience*. Cambridge: Cambridge University Press, 1973.

Terleckyi, N. E. Direct and indirect effects of industrial research and development on the productivity growth of industries. In Kendrick J. W. and Vaccara, B. N. (eds.) *New developments in productivity measurement and analysis*. Chicago: University of Chicago Press, 1980.

Teubal, M. and Steinmueller, E. Government policy, innovation, and economic growth. *Research policy*, 1982, 271–87.

Toren, N. and Galai, D. The determinants of the potential effectiveness of government-supported industrial research institutes. *Research policy*, 1978, 7, 362–382.

UNIDO (Dasgupta, P., Sen, A. and Marglin, S.) Guidelines for project evaluation. New York: United Nations, 1972.

Weitzman, M. L. Efficient incentive contracts. *The quarterly journal of economics*, 1980, 719–730.

Wright, B. D. The economics of invention incentives: Patents, prizes and research contracts. *American economic review*, 1983, 691–707.